The Mentality of Apes

International Library of Psychology, Philosophy and Scientific Method

Editor: C. K. Ogden

A Catalogue of books already published in the *International Library of Psychology, Philosophy and Scientific Method* will be found at the end of this volume.

The Mentality of Apes

Wolfgang Köhler

Translated from the second revised edition by
Ella Winter

Routledge & Kegan Paul

London and Boston

First published in 1925
Second edition (revised and reset) 1927
Reprinted 1948, 1973
by Routledge & Kegan Paul Ltd.
Broadway House, 68-74 Carter Lane,
London EC4V 5EL and
9 Park Street,
Boston, Mass. 02108, U.S.A.

Printed in Great Britain by
Redwood Press Limited
Trowbridge, Wiltshire

ISBN 0 7100 7525 1

CONTENTS

		PAGE
INTRODUCTION	I
I. ROUNDABOUT METHODS	II
II. THE USE OF IMPLEMENTS . . .	25
III. THE USE OF IMPLEMENTS (*cont.*) HANDLING OF OBJECTS	67
IV. THE MAKING OF IMPLEMENTS . . .	99
V. THE MAKING OF IMPLEMENTS (*cont.*) BUILDING .	135
VI. DETOURS WITH INTERMEDIATE OBJECTIVES .	173
VII. "CHANGE" AND "IMITATION" . . .	185
VIII. THE HANDLING OF FORMS . . .	225
CONCLUSION	265
APPENDIX : SOME CONTRIBUTIONS TO THE PSYCHOLOGY OF CHIMPANZEES . .	271
INDEX	331

71974

LIST OF ILLUSTRATIONS

PLATE

I. NUEVA FIVE DAYS BEFORE HER DEATH . *Frontispiece*

PAGE

II. CHICA ON THE JUMPING-STICK . . . 72

III. SULTAN MAKING A DOUBLE-STICK . . 128

IV. GRANDE ON AN INSECURE CONSTRUCTION . 138

V. GRANDE ACHIEVES A FOUR-STORY STRUCTURE . 142

VI. CHICA BEATING DOWN HER OBJECTIVE WITH A
POLE 146

VII. KONSUL, GRANDE, SULTAN AND CHICA BUILDING 168

VIII. WOODEN BANANA-BASKET FULL AND EMPTY . 326

IX. BANANAS AND STONE 328

PREFACE

THIS book contains the results of my studies in the intelligence of Apes at the Anthropoid Station in Tenerife from the years 1913-1917. The original, which appeared in 1917, has been out of print for some time. I have taken this opportunity of making a few changes in the critical and explanatory sections, and have added as an Appendix some general considerations on the Psychology of Chimpanzees.

With various recent books and essays on the subject I shall have an opportunity of dealing in a further contribution to the subject not yet completed.

W. KÖHLER

BERLIN, *October* 1924.

TRANSLATOR'S NOTE

THE terminology used in this translation was agreed upon after detailed discussion between author and translator. Often the only method possible of covering all the implications of the German terminology was to use several different English terms : as has been done with such words as *Einsicht, Umweg, Gestalt, das Zueinander von Gestalten,* etc. Attention has been drawn to most of these cases in translator's footnotes.

The paragraphs in square brackets, correspond to sections printed in small type in the original, and denote supplementary explanations or digressions.

ELLA WINTER.

THE MENTALITY OF APES

INTRODUCTION

1. Two sets of interests lead us to test the intelligence of the higher apes. We are aware that it is a question of beings which in many ways are nearer to man than to the other ape species ; in particular it has been shown that the chemistry of their bodies, in so far as it may be perceived in the quality of the blood, and the structure of their most highly-developed organ, the brain, are more closely related to the chemistry of the human body and human brain-structure than to the chemical nature of the lower apes and *their* brain development. These beings show so many human traits in their " everyday " behaviour that the question naturally arises whether they do not behave with intelligence and insight under conditions which require such behaviour. This question expresses the first, one may say, naïve, interest in the intellectual capacity of animals. We wished to ascertain the degree of relationship between anthropoid apes and man in a field which seems to us particularly important, but on which we have as yet little information.

The second aim is theoretical. Even assuming that the anthropoid ape behaves intelligently in the sense in which the word is applied to man, there is yet from the very start no doubt that he remains in this respect far behind man, becoming perplexed and making mistakes in relatively simple situations ; but it is precisely for this reason that we may, under the simplest conditions, gain knowledge of the nature of intelligent acts. The human adult seldom performs

1

for the *first* time in his life tasks involving intelligence of
so simple a nature that they can be easily investigated ;
and when in more complicated tasks adult men really find
a solution, they can only with difficulty observe their own
procedure. So one may be allowed the expectation that in
the intelligent performances of anthropoid apes we may
see once more in their plastic state processes with which we
have become so familiar that we can no longer immediately
recognize their original form : but which, because of their
very simplicity, we should treat as the logical starting-point
of theoretical speculation.

As all the emphasis in the following investigations is laid
on the first question, the doubt may be expressed whether
it does not take for granted a particular solution of the
problems treated under the second. One might say that
the question whether intelligent behaviour exists among
anthropoid apes can be discussed only after recognizing
the theoretical necessity of distinguishing between intelligent
behaviour and behaviour of any other kind ; and that,
since association psychology, in particular, claims to derive
from one single principle all behaviour which would come
under consideration here, up to the highest level, even that
attained by human beings, a theoretical point of view is
already assumed by the formulation of problem 1 ; and
one which is antagonistic to association psychology.

This is a misconception. There is probably no association psychologist who does not, in his own unprejudiced
observations, distinguish, and, to a certain extent, contrast,
unintelligent and intelligent behaviour. For what is association psychology but the theory that one can trace back
to the phenomena of a generally-known simple association
type even those occurrences which, to unbiassed observation,
do not at first seem corresponding to that type, most of all
the so-called intelligent performances ? In short, it is just

these differences which are the starting-point of a strict association psychology ; it is they which need to be theoretically accounted for ; they are well known to the association psychologist. Thus, for instance, we find a radical representative of this school (Thorndike) stating the conclusion, drawn from experiments on dogs and cats : " I failed to find any act that even *seemed* due to reasoning." To anyone who can formulate his results thus, other behaviour must have seemed to be intelligent ; he is already acquainted with the contrast in his observations, say of human beings, even if he discards it afterwards in theory.

Accordingly, if we are to inquire whether the anthropoid ape behaves intelligently, this problem can for the present be treated quite independently of theoretical assumptions, particularly those for or against the association theory. It is true that it then becomes somewhat indefinite ; we are not to inquire whether anthropoid apes show something well defined, but whether their behaviour approximates to a type rather superficially known by experience, and which we call "intelligent "[1] in contrast to other behaviour— especially in animals. But in proceeding thus, we are only dealing according to the nature of the subject ; for clear definitions have no place at the beginning of sciences founded on experience ; it is only as we advance towards results that we can mark our progress by the formulation of definitions.

Moreover, the type of human and, perhaps, animal behaviour to which the first question animadverts is not quite indefinite, even without a theory. As experience shows, we do not speak of behaviour as being intelligent, when human beings or animals attain their objective by a direct unquestionable route which clearly arises naturally out of their organization. But we tend to speak of " intelligence " when, circumstances having blocked the obvious course, the human being or animal

[1] See foot-note, p. 219.

takes a roundabout path, so meeting the situation. In unexpressed agreement with this, nearly all those observers who heretofore have sought to solve the problem of animal intelligence, have done so by watching animals in just such predicaments. Since animals below the stage of development of anthropoid apes give, in general, negative results, there has arisen out of these experiments the view widely held at present, i.e., that there is very little intelligent behaviour in animals. Only a small number of such experiments have been carried out on anthropoid apes, and they have not yet produced any very definite results. All the experiments described in the following pages are of one and the same kind : the experimenter sets up a situation in which the direct path to the objective is blocked, but a roundabout way left open. The animal is introduced into this situation, which can, potentially, be wholly surveyed. So we can see of what levels of behaviour it is capable, and, particularly, whether it can solve the problem in the possible " roundabout " way.

2. The experiments were at first applied to chimpanzees only, with the exception of a few cases taken for comparison, in which human beings, a dog, and hens were observed.

Seven of the animals belonged to the old branch of the anthropoid station which the Prussian Academy of Science maintained in Tenerife from 1912 to 1920. Of these seven the oldest, an adult female, was named Tschego, because of several characteristics which made us, perhaps wrongly, consider her a member of the Tschego species. (We are yet far from possessing a clear and systematized classification of the varieties of the chimpanzee.) The oldest of the smaller animals, called Grande, differed considerably in several respects from its comrades. But as the differences concern its general character rather than the behaviour investigated in the intelligence tests, a detailed description of them would be out of place here. The other five, two males (Sultan and

Konsul), and three females (Tercera, Rana, and Chica), were of the usual chimpanzee type.

To the seven animals mentioned, two others were added later, both of which led to valuable observations, but both of which, to our regret, soon died. I shall briefly describe them in order to give an impression of the completely different " personalities " which exist among chimpanzees.

Nueva, a female ape, about the same age as the other little animals (four to seven years at the time of the majority of our experiments), differed from them bodily in her extraordinarily broad ugly face and an obviously pathological sparsity of hair on her unhealthy skin. But her ugliness was completely offset by a nature so mild and friendly, of such naïve confidence and quiet clarity as never fell to our lot to meet with in a chimpanzee before or after. Her childlike attachment we found to some extent in other animals when they were ill, and perhaps many of Nueva's good qualities can be explained by the fact that, from the beginning, she was the prey of a slowly-advancing disease ; chimpanzees, on the whole, can do with a little suppression. We were particularly impressed by the way she would play for hours, quite contentedly, with the simplest toys. Unfortunately the others tended to become lazy if they were not given any particular employment, or if they were not quarrelling, or inspecting each other's bodies. If a number of healthy children are left together all the time, without any particular occupation, the effect will not be in the line of a discreet, though playful activity either. Nueva had been kept alone for many months. One must, however, not assume that the pleasant qualities of this animal were due to earlier educational influences. Unfortunately, education does not seem able to transform a naturally mischievous and wanton chimpanzee into an amiable being ; moreover, Nueva was not " brought up " in the nursery sense ; on the contrary, she showed that she was not used to being corrected at all. She regularly ate her excretions, and was first astonished and then extremely indignant when we took measures against this habit. On the second day of her stay at the station, the keeper threatened her, during this proceeding, with a little stick, but she did not understand the meaning of the stick, and wanted to play with it. If food which she had, with complete naïveté, appropriated somewhere, was taken away from her she would bite, in her sudden rage, immediately ; she was as yet without any inhibitions towards man ; in fact, she showed herself completely naïve, and was, without doubt, less " cultured " than the station animals.

The male, Koko, judged to be about three years of age, was a type of chimpanzee not uncommonly met with : above his drum-taut stomach a pretty face with neatly parted hair, a pointed chin, and prominent eyes which seemed always discontentedly asking for something, giving the little fellow a native expression of sauciness. A large part of his existence was, in fact, spent in a kind of chronic indignation, either because there was not enough to eat, or because the children came too near him, or because someone who had just been with him left him again, or finally, because he could not remember to-day how he had solved a similar test yester-day. He would not complain ; he would merely be indignant. Usually this mood was manifested by loud pommelling on the floor with both fists, and an agitated hopping up and down in one spot ; in cases of great rage by glottal cramp-attacks which passed over quickly. (These we noticed also in other chimpanzees when they had attacks of rage, and very rarely in manifestations of joy.) Before such attacks, and in cases of minor excitement, he would utter a continual staccato ŏ in that irregular characteristic rhythm which one hears from a slow-firing line of soldiers. In his angrily-uttered demands, and his wild indignation if they were not immediately satisfied, Koko resembled another egoist *par excellence*, Sultan. Luckily—and perhaps that is no accident —Koko was, at the same time, just as gifted as Sultan.

These are only two chimpanzees. For one who has seen Koko and Nueva alive, there is no doubt that in their own way they were as much unlike as two human children with fundamentally different characters, and one can set up as a general maxim that observations of one chimpanzee should never be considered typical for all of this species of animal. The experiments we describe in the following show that there are just as great individual differences in the intellectual field.

Practically all the observations were made in the first six months of 1914.[1] They were frequently repeated later, but only a few additional experiments and repetitions (dating from the spring of 1916) are incorporated in this re¬ort, as, in general, the behaviour observed the first time was repeated ; in any case, no important corrections had to be made in the earlier results.

3. Experiments of the kind described above may make very different calls upon the animals to be tested, according to the situation in which they are put. In order to discover, even roughly, the zone of difficulty within which the testing of chimpanzees will be of any use, Mr. E. Teuber and I gave them a problem which seemed to us difficult, but not impossible, of solution for a chimpanzee. How Sultan behaved in this test should be sketched here as a preliminary example.

A long thin string is tied to the handle of a little open basket containing fruit ; an iron ring is hung in the wire-roof of the animals' playground through which the string is

---

[1] That is, they were made *before* the chimpanzees underwent optical examination. (Cf. these in the *Abh. d. Kgl. Preuss. Akd. d. Wiss*, 1915, Phys.-Math. Section No. 3.)

pulled till the basket hangs about two metres above the
ground; the free end of the string, tied into a wide open
loop, is laid over the stump of a tree-branch about three
metres away from the basket, and about the same height
from the ground; the string forms an acute angle—the bend
being at the iron ring (cf. Fig. 1). Sultan, who has not seen
the preparations, but who knows the basket well from his
feeding-times, is let into the playground while the observer
takes his place outside the bars. The animal looks at the

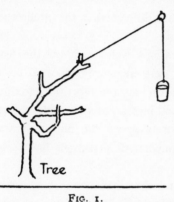

hanging basket, and soon
shows signs of lively agitation (on account of his
unwonted isolation), thun-
ders, in true chimpanzee
style, with his feet against
a wooden wall, and tries to
get into touch with the
other animals at the win-
dows of the ape-house and
wherever there is an outlook,
and also with the observer

FIG. 1.

at the bars; but the animals are out of sight, and the observer
remains indifferent. After a time, Sultan suddenly makes
for the tree, climbs quickly up to the loop, stops a moment,
then, watching the basket, pulls the string till the basket
bumps against the ring (at the roof), lets it go again, pulls
a second time more vigorously so that the basket turns over,
and a banana falls out. He comes down, takes the fruit,
gets up again, and now pulls so violently that the string
breaks, and the whole basket falls. He clambers down,
takes the basket, and goes off to eat the fruit.

Three days later, the same experiment is repeated, except
that the loop is replaced by an iron ring at the end of the rope,
and the ring, instead of being put over the branch, is hung

on a nail driven into a scaffolding (used for the animals' gymnastics). Sultan now shows himself free from all doubt, looks up at the basket an instant, goes straight up to the scaffolding, climbs it, pulls once at the cord, and lets it slip back, pulls again with all his might so that the cord breaks, then he clambers down, and fetches his fruit.

The best solution of the problem which could be expected would be that the animal should take the loop or iron ring off the branch or nail and simply let the basket drop, etc. The actual behaviour of the animal shows plainly that the hub of the situation, i.e., the rope connexion, is grasped as a matter of course, but the further course of action for the experiment is not very clear. The best solution is not even indicated. One cannot tell just why. Did Sultan perhaps not see the loose fixing of the loop to the branch or ring to the nail ? If he had noticed it, would he have been able to solve it ? Would he in any case expect the basket to fall to the ground if this fastening were loosened ? Or does the difficulty lie in the fact that the basket would fall to the ground, and not straight into Sultan's hands ? For we cannot even know whether Sultan really pulled at the cord to break it, and thus bring the basket to earth. So we have performed one experiment which, for a beginning, contains conditions too complicated to teach us much, and, therefore, we see the necessity of beginning the next examinations with elementary problems in which, if possible, the animals' conduct can have one meaning only.

I

ROUNDABOUT METHODS[1]

WHEN any of those higher animals, which make use of vision, notice food (or any other objective) somewhere in their field of vision, they tend—so long as no complications arise— to go after it in a straight line. We may assume that this conduct is determined without any previous experience, providing only that their nerves and muscles are mature enough to carry it out.

Thus, if the principle of experimentation mentioned in the introduction is to be applied in a very simple form, we may use the phrases " direct way " and " roundabout way " quite literally, and set a problem which, in place of the biologically-determined direct way, necessitates a complicated geometry of movement towards the objective. The direct way is blocked in such a manner that the obstacle is quite easily seen ; the objective remains in an otherwise free field, but is attainable only by a roundabout route. First it is assumed that the objective, the obstruction, and also the total field of possible roundabout routes are in plain sight ; if the obstruction be given various forms, there will develop also a variety of approaches to the objective, and perhaps, at the same time, variations in the difficulties which such a situation contains for the animal.

This test which, on nearer investigation, appears to be the simplest and, in some respects, fundamental for theoretical

[1] Also called *detours, roundabout ways, paths or routes, circuitous routes* and *indirect ways* in this book.   No one English word quite covers the meaning of " Uwege " [Tr. Note.].

problems, will, in chimpanzees from four to seven years of age and in the form described, yield no results which cannot be observed in their ordinary behaviour. Chimpanzees will get round any obstruction lying between them and their objective, if they have sufficient view of the space in which lie the possible detours. The path may lie across flat ground, or over trees and scaffolding, or even up under a roof as long as they can grab hold of something. Thus in experiments to be described later, in which the objective

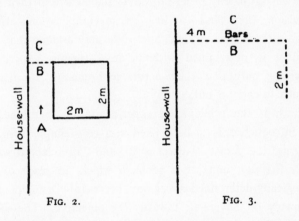

FIG. 2.   FIG. 3.

hung from the wire-roof of their playground, the first attempt at solution often consisted in their climbing to the roof at the first available point, and thence arriving at the hanging cord. It required strict vigilance to eliminate from the programme this and other detours which only climbers like chimpanzees, and among them only the real acrobats, like Chica, would hit upon. For it must not be assumed that even in bodily dexterity chimpanzees are all alike. One sees the animals twisting, bending, and turning their bodies with equal facility according to the shape of an entrance ; but no one expects a chimpanzee to remain helpless before a horizontal opening in a wall, on the other side of which his objective lies, and so it makes no impression at all on us when he makes as horizontal a shape as he can of himself, and thus slips through. It is only when round-about methods are tried on the lower animals, and when you see even chimpanzees undecided, nay, perplexed to the point of helplessness, by a seemingly minor modification of the problem—it is only then you realize that circuitous methods cannot in general be considered usual and matter-of-course conduct.[1] But, as chimpanzees do not give us the impression of any particular insight when they take a round-about route (at any rate in the form so far discussed) no further explanation is here required, because of the non-theoretical form of our problem.

Meanwhile, however, in the simplest experiments of the roundabout type, observation is so easy that a description of such tests performed on other animals is advisable. Taking such a simple case as an example, one becomes aware of a factor which occurs over and over again in all difficult experiments with chimpanzees, and will be more easily observed there after it has become familiar here. Therefore the following examples are quoted.

Near the wall of a house, a square piece of ground is fenced off so that one side, one metre from the house, is parallel to it, and forms with it a passage two metres long ; one end of this passage is cut off by a railing. A mature Canary Isle bitch is brought into this blind alley from direction A (cf. Fig. 2), to B, where she is kept occupied with food, her face towards the railings. When the food is nearly gone, more is put down at the spot C, on the other side of the rail ; the bitch sees it, seems to hesitate a moment, then quickly turns at an angle of 180° and is already on the run in a smooth curve, without any interruption, out of the blind alley, round the fence to the new food.

The same dog, on another occasion, behaved at first in

[1] Cf. the last section of this book, p. 226, seqq.

the same way. It was standing at B near a wire fence (con structed as in Fig. 3) over which food was thrown to some distance ; the bitch at once dashed out to it, describing a wide bend. It is worth noting that when, on repeating this experiment, the food was not thrown far out, but was dropped just outside the fence, so that it lay directly in front of her, separated only by the wire, she stood seemingly helpless, as if the very nearness of the object and her concentration upon it (brought about by her sense of smell) blocked the " idea " of the wide circle round the fence ; she pushed again and again with her nose at the wire fence, and did not budge from the spot.

A little girl of one year and three months, who had learned to walk alone a few weeks before, was brought into a blind alley, set up *ad hoc* (two metres long, and one and a half wide), and, on the other side of the partition, some attractive object was put before her eyes ; first she pushed towards the object, i.e., against the partition, then looked round slowly, let her eyes run along the blind alley, suddenly laughed joyfully, and in one movement was off on a trot round the corner to the objective.

In similar experiments with hens, one sees that a roundabout way is not taken as a matter of course, but is quite an achievement ; hens, in situations which are much less roundabout than those already described, have been quite helpless ; they keep rushing up against the obstruction when they see their objective in front of them through a wire fence, rush from one side to the other all a-fluster, and do not fare better, even when they are familiar with the obstruction (or the fence) and the greater part of the circuitous route, as, for instance, round the little door of their place and through the opening corresponding to it. Different hens do not behave in the same way, and, if the detour is shortened while they are still pushing against the obstacle, it can easily be observed how first one, then another, and so on, stops running up against the obstruction, and runs quickly round the curve ; but some particularly ungifted specimens keep on running up against the fence a long while even in the simplest predicaments. The difference is very plain too, when one notices in cases of longer roundabout routes to what an extent chance must help to solve the problem. In their

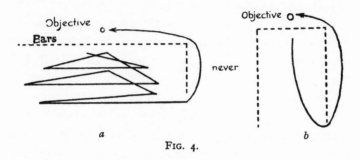

FIG. 4.

oscillations in front of the objective, the hens now and then run into places from which the circuitous route is shorter ; but this easing-up brought about by chance will have a very different effect on different animals : one will suddenly rush out in a closed circle, another will still zigzag helplessly to and fro in the " wrong " direction. All the hens which I observed thus managed to achieve only very " straight " roundabout ways (cf. Fig. 4a in contrast to 4b). Apparently the possible detour must not begin with the direction leading away from the objective (cf. as against this the behaviour of the child and the dog above).

It therefore follows that for those processes which form the basis of this small achievement, variations in the geometrical circumstances are of the greatest importance.[1] The influence of these circumstances will more than once be striking in the

[1] In what way they are dependent can be more exactly determined, and when this is known, definite conditions will exist for every theory concerning the experiment.

case of the anthropoids, in what are, for them, much harder tasks.

As chance can bring the animals into more favourable spots, it will also occasionally happen that a series of pure coincidences will lead them from their starting-point right up to the objective, or at least to points from which a straight path leads to the objective. This holds in all intelligence tests (at least in principle : for the more complex the problem to be solved, the less likelihood is there that it will be solved wholly by chance) ; and, therefore, we have not only to answer the question whether an animal in an experiment will find the roundabout way (in the wider meaning of the word) at all, we have to add the limiting condition, that results of chance shall be excluded. Now (if we take as examples these experiments in roundabout ways—in the narrower sense) since approximately the same path must be followed by the animal, whether as the result of a succession of accidents, or of a real solution of the problem, the objection will arise, that one cannot distinguish between these two possibilities. It is of great importance for what follows and for the psychology of the higher animals in general, that one should not allow oneself to be confused by such apparently " pat " but, in reality, false, considerations. *Observation, which alone may be admitted here, shows that there is in general a rough difference in form between genuine achievement and the imitations of accident,* and no one who has performed similar experiments on animals (or children) will be able to disregard this difference. The genuine achievement takes place as a single continuous occurrence, a unity, as it were, in space as well as in time ; in our example as one continuous run, without a second's stop, right up to the objective. A successful chance solution consists of an agglomeration of separate movements, which start, finish, start again, remain independent of one another in direction and speed, and only in a geometrical summation start at the starting-point, and finish at the objective. The experiments on hens illustrate the contrast in a particularly striking way, when the animal, under pressure of the desire to reach the objective, first flies about uncertainly (in zigzag movements which are shown in Fig. 4*a* but in not nearly great enough confusion), and then, if one of these zigzags leads to a favourable place, suddenly rushes along the curve in one single unbroken run. Here, the first part of the possible path is swallowed up in confused zigzagging, all the rest is " genuine "—the one type of behaviour succeeding the other so abruptly that no one could mistake the difference in the two kinds of movements.

If the experiment has not been made often, there is the additional fact that the moment in which a true solution is struck is generally sharply marked in the behaviour of the animal (or the child) by a kind of jerk : the dog stops, then suddenly turns completely round (180°), etc., the child looks about, suddenly its face lights up, and so forth. Thus the characteristic smoothness of the true solution is made more striking by a discontinuity at its beginning.

I must explicitly warn my readers against the mistake of thinking that I am implying any supernatural mode of interpreting behaviour : any practised person can observe this, not only in experiments on animals, but in all others. Similar considerations have to be taken into account often enough outside the animal world. Thus, wandering earth-currents, and other rapidly-alternating fortuitous influences, deflect the thread of a badly set-up electrical measuring instrument irregularly to and fro on the scale ; but should the thread move constantly to a certain scale division, no physicist would mistake the evident difference, and its meaning. In observing the Brownian movement any experimental error which causes the introduction of a regular movement into one which is normally irregular would at once be detected, and so forth. Later on, more will be said about this matter, the importance of which does not concern method alone.

[Experiments in roundabout ways of the kind described must not be confused with two other experimental methods : 1. " Frogs without brain and mid-brain still get out of the way of obstacles " (Nagel, *Physiol. des Menschen*, IV, I, p. 4 ; A. Tschermak). Thus the animals move automatically out of a line of motion which would bring them into collision with an obstacle. Does it follow that the same frogs would automatically take a long way round an obstacle up to an objective ? Obviously not. The main point in our experiment does not arise at all in the frog experiment. 2. American animal psychology makes animals (or people) seek the way out of mazes, over the whole of which there is no general survey from any point inside ; the first time they get out is, therefore, necessarily a matter of chance, and so, for these scientists, the chief question is how the experience gained in such circumstances can be applied in further tests. In intelligence tests of the nature of our roundabout-way experiments, everything depends upon the situation being surveyable by the subject from the outset.]

I made the experiment more difficult for chimpanzees, in the following way : The objective hangs in a basket from the wire-roof and cannot be reached from the ground ; the basket contains also several heavy stones, so that one push of the string and basket will make the whole swing for some little time ; the swing is so arranged that the longest sideways movement of the basket makes it nearly reach a scaffolding. Thus the roundabout way is easily recognizable, and available, but only for a few moments.—(19.1.14)—As soon as the basket is swinging, Chica, Grande, and Tercera are let in upon the scene.[1] Grande leaps for the basket from the ground, and misses it. Chica who, in the meantime, has quietly surveyed the situation, suddenly runs towards the scaffolding, waits with outstretched arms for the basket, and catches it. The experiment lasted about a minute.[2]

Repetitions with other animals (Rana, Koko) also went so smoothly and quickly that one can probably infer that every chimpanzee can solve this problem. Grande, who had seen Chica's solution, duplicated it on an immediate repetition of the test. Judging by everything that happened later, there is no doubt that example is not absolutely necessary, and that, always slower than the others, Grande would, after a little while, have seen the roundabout route of herself.

Sultan, who was not present at these experiments, was tested with the same swing (20.1), but this time, before he saw it, the basket was set swinging in a circle which brought it at regular speed past a beam; the circular swing and the regular speed doubtless made this experiment a little harder. Sultan looked up for a second, and followed the basket with his eyes; when he saw it swinging past the beam, he was up there at once, awaiting it.

[1] In the first few days these animals were far too timid to permit of the isolation of any one of them for experimentation; this circumstance caused the very greatest difficulties, and even after six months it was still impossible to test Chica alòne. Usually in such cases I gave Tercera or Konsul as companions; for they were not much use anyhow on account of shyness or laziness; but other subjects of experiment were sometimes similarly wasted.

[2] In this book I give either no times at all or the approximate time in those instances where it bears on the subject. In general the duration of an experiment depends on so many accidental and changing circumstances (e.g. futile attempts at solution, lack of interest, depression on account of failure or isolation, etc.) that measures of time would only give the *semblance* of a quantitative method. The time-data in any of these experiments can always be judged or estimated from the description, as far as it is important for our purposes. Whether an interval of indifference or complaining, as often occurred, lasted three minutes, i.e. perhaps ten times as long as the actual time of solution, or half an hour, perhaps a thousand times the length of that, does not matter at all. In most cases the solution itself would make up any fraction one liked of the measured " duration of the experiment ".

In experiments such as these, it does not matter at all whether the point which the swing approaches remains the same in successive experiments or not; and neither does it matter whether the vantage-point is a wall, a tree, a scaffolding, or anything else. If variations of this sort are introduced, the animal does not climb up to the spot at which it was successful before; it clambers with complete certainty to the right place for the new situation. In experiments as simple as this I never saw this rule broken, but in harder tasks, mistakes involving stupid repetitions did occur.

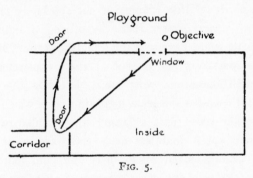

FIG. 5.

The experiment is considerably more difficult when a part of the problem, if possible the greater part, is not visible from the starting-point, but is known only " from experience."

One room of the monkey-house has a very high window, with wooden shutters, that looks out on the playground. The playground is reached from the room by a door, which leads into the corridor, a short part of this corridor, and a door opening on to the playground (cf. Fig. 5). All the parts mentioned are well known to the chimpanzees, but animals in that room can *see* only the interior. (6.3)—I take Sultan with me from another room of the monkey-house, where he was playing with the others, lead him across the corridor into that room, lean the door to behind us, go with him to the window, open the wooden shutter a little, throw a banana out, so that Sultan can see it disappear through the window, but, on account of its height, does not see it fall, and then quickly close the shutter again (Sultan can only have seen a little of the wire-roof outside). When I turn round Sultan is already on the way, pushes the door open, vanishes down the corridor, and is then to be heard at the second door, and immediately after in front of the window. I find him outside, eagerly searching underneath the window ; the banana has happened to fall into the dark crack between two boxes. Thus not to be able to see the place where the objective is, and the greater part of the possible indirect way to it, does not seem to hinder a solution ; if the lay of the land be known beforehand, the indirect circuit through it can be apprehended with ease.

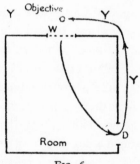

Fig. 6.

In a very similar experiment with the bitch already mentioned, she managed the same manœuvre. From the yard which runs straight and unencumbered around the house, one steps through the door D into a room with its window W looking out on to the yard Y (cf. Fig. 6) ; the bitch, who is acquainted with the room and the yard from former visits—she does not belong to the house—is brought through the door D into the room, and is tempted, with food, to the open window ; from here she can see only the tops of distant trees, not the yard itself. The food is thrown out, and the window at once shut. The dog jumps once against the window-pane, then stands a moment, her head raised towards the window, looks a second at the observer, when all at once she wags her tail a few times, with one leap whirls round 180°, dashes out of the door, and runs round outside, till she is underneath the window, where she finds the food immediately.[1]

[Thorndike tested large numbers of dogs and cats in order to see what there is in the wonder-stories that are told about these domestic pets. The result was very unfavourable to the animals, and Thorndike came to the conclusion that, so far from " reasoning ", they do not even associate images with perception, as humans do, but remain limited chiefly to the experiential linking of mere " impulses " with perceptions. This investigation did what was necessary in a negative way at the time, but, as is now being shown (also in America), it went a little too far. The tests were based upon those animal stories, and consequently were made so difficult that the result *was bound* to fall out badly ; under the influence of the animals' failures in these tests, Thorndike then drew generalizations about their capacities, which do not follow from those difficult experiments. However stupid a dog may seem compared to a chimpanzee, we suggest that in such simple cases as have just been described, a closer investigation would be desirable.

Regarding their principle, I must make a further objection to Thorndike's experiments. They were designed as *intelligence tests* of the same type as our own (insight or not ?)[2], and ought, therefore, to have conformed to the same general conditions, and, above all, to have been arranged so as to be completely *visible* to the animals. For if essential portions of the experimental apparatus cannot be seen by the animals, how can they use their intelligence faculties in tackling the situation ? It is somewhat astonishing to find that

---

[1] Somewhat different experiments in detours were made by Thorndike (cf. the work quoted below) and Hobhouse (*Mind in Evolution*, London, 1901, p. 223 seqq.). I must add that the bitch was not brought through the door from the window-side of the house, so that behind her she can only have had a scent-trail as far as the door ; in any case her sense of smell was not observed to have played any part at all.

[2] See foot-note, p. 219.

(in Thorndike's experiments) cats and dogs were frequently placed in cages containing the *extreme end* only of one or the other mechanism, or allowing a view of ropes or other parts of the mechanism, but from which a survey over the *whole* arrangement was not possible. The task for the animal was to let itself out of the cage by pulling or pressing the accessible part of the mechanism ; then—the cage door would open of itself. Thorndike also gives an account of experiments in which the animals were let out of their cages if they scratched or licked themselves. He contrasts these experiments with those involving the employment of any mechanical contrivance, .as the former apparently imply no direct connexion between cause and effect ; but the causation is far from apparent even in the mechanistic experiments.

In the case of the latter, there are at least various *component parts* which can be treated with some amount of insight, and it is of the highest significance to know whether animals react differently to experimental situations which involve a partial possibility of intelligent behaviour than they do to such as involve none—for the difference, if any, is obviously crucial.

The result of these experiments tends to show that prolonged " learning " is necessary before the right action develops, in *both* sets—as the " experiments with a mechanism " were far too difficult, and, in many cases, could not be wholly surveyed either. But when once the animals have mastered both procedures, a noticeable difference is shown : " In all these cases "—of the meaningless type—" there is a noticeable tendency . . . to diminish the required action, till it becomes a mere vestige of a lick or scratch "—and more especially—" if sometimes you do not let the cat out after this feeble reaction, it does not at once repeat the movement, as it would do if it depressed a thumb-piece, for instance, without success in getting the door open."[1]

Thorndike merely states that he cannot give a reason for the difference of result in the two types of experiment. As these results are among the most interesting which he has obtained—though scarcely what we might expect from his theory—we can only regret that he has not probed further.]

[1] *Animal Intelligence*, New York, 1911, p. 48.

II

THE USE OF IMPLEMENTS

THE situation is made more difficult. The objective in view cannot be reached by making a detour, nor can the body of the animal be adapted to the shape of its surroundings, and thus reach the objective. If the connexion between animal and objective is to be established, it can only be through, and by means of, a third body. This cautious manner of expression is necessary in exposition here ; it is only in the case of certain forms of this procedure with third bodies that it is permissible to say that the objective " is secured by means of a tool or implement." This description does not give a correct idea of some ways of overcoming the distance between animal and objective by means of a third object.[1]

If the field of action contains third objects or bodies appropriate for overcoming the critical distance between the animal and its objective the question is : How far is a chimpanzee capable of making use of such objects in the drive to reach the objective ?

(1)

The problem is most easily solved when the distance (between goal and animal) is already virtually overcome,

[1] Moreover it is advisable in this whole field to replace stereotyped terms like " use of implements " and " imitation ", by phrases which shall be so far as possible exactly descriptive of the animal's actions. Those words are clichés which conceal the most important questions under a mask of commonplace. It is far more illuminating to let the animal's behaviour determine one's words, but certainly often more difficult, as there are often no appropriate words in our languages.

i.e. when the implement is already "placed" in relation to the goal. The connexion can either be used or ignored, the implement appearing of no interest, and the animal remaining helpless.

In the introduction we saw that Sultan is master of such a situation, although it is not of the simplest type, and the connexion between implement and objective only becomes of account when he climbs a tree. If the experiment is so far simplified that, for instance, a string fastened to the objective is within reach of the animal, then the chimpanzee will almost always solve the problem immediately.

The experiment was made with Nueva on the sixth day of her sojourn at the station (14.3). The objective was at a distance of over one metre from the bars of her cage, and a soft straw was tied to it, lying across the intervening space, which was otherwise empty, right up to the bars. Neuva had hardly seen the objective before she seized the straw and carefully pulled the prize towards her.

Koko had been for five days at the station : (13.7). He was tied by his collar and chain to a tree, so that his range of movement was limited. The objective had been placed on the ground outside this periphery, and a string fastened to the objective was within Koko's reach. He had not seen the preparations. When his attention was drawn to the objective, he glanced at it for a moment, and then turned away ; he was again directed towards the edible, seized the string, and drew the objective towards him—only to throw it away after a brief examination : it was not good enough.

Two of the chimpanzees, Tschego and Konsul, had given the same positive results in this experiment (14.2), although in their cases the rope measured *three* metres. The other animals were all confronted with the same problem, in more complicated experiments, and not one of them ever hesitated to use the rope attached to the objective. And the pulling was always " with an eye on the objective " in the strict sense of the words ; one glance at the objective, and the animal began to pull the rope, gazing, not at the rope, but at the distant objective. There can be no possibility that the object of attention was the rope, which for some reason or other was drawn towards the animal.

[*Variation of this Experiment.*

The objective is placed inside a basket ; a rope is attached to the handle and leads up to the barred window of the room in which the chimpanzee is kept : the basket is always hoisted up by the rope.]

In a similar situation, a dog would be able to help himself by means of his teeth or forepaws ; but the afore-mentioned bitch did not even attempt this simple method of self-help, and paid no attention to the string which was lying just under her nose—whilst at the same time she showed the liveliest interest in the distant objective. Dogs, and probably, for instance, horses as well, unless they made sudden lucky movements or received indications from outside themselves, might easily starve to death in these circumstances which offer hardly any difficulty to human beings—or to chimpanzees.

The achievement of the anthropoids, however, deserves more careful consideration. To this end, the circumstances were intentionally somewhat confused : (11.6.1914). The objective, tied to a string, lies on the ground on the further side of the bars, and three more strings, besides the " right " one, run from the approximate direction of the objective, crossing the " right " string and each other. Their ends lie near the bars (cf. Fig. 7*a*). An adult human being, with only a slight degree of attention, can perceive at once which string is the right one. Sultan is led to the bars, glances out, and then pulls in rapid succession, two of the wrong strings, and then the right one (the order is indicated by the numerals in Fig. 7a).

The field is clearer if only two strings, one right and one wrong, run towards the objective, without necessarily crossing. The appended sketch 7 shows the result in four instances *b* to *e*. (14.6). The distance between the ape and the objective amounts to about one metre and the " wrong " string has one end placed about five centimetres from the objective. These experiments do not permit us to form any conclusion as to Sultan's ability to recognize the " right " string after careful observation, for Sultan never gives himself time to make the effort such attention requires, but simply grips something and pulls—and only twice at once hits on the right string. His mistakes can scarcely be fortuitous ; in the course of *five* experiments, he gave the first pull four times to the string which appeared to reach the objective by the shortest distance from the bars.

[There is possibly also a tendency to give the preference to strings lying to the *right* ; this would be susceptible to a simple *motor* explanation ; for Sultan always takes up his position directly opposite to the objective, and, on all occasions which require the slightest degree of skill, uses his *right* hand.]

When only one string lies with its further end in the neighbourhood of, but unattached to, the objective, everything depends on *how* near it lies. In one of this set of experiments the objective was three metres from the bars, and the end of the string fifteen centimetres from the objective. Sultan glanced cursorily at the end of the string near the objective, but did not move ; a few seconds later he began to pull the string, without, however, in the least noticing the objective ; all his attention was on the string, and he began to play with it ; yet he was in the best of appetites, as the offer of food showed. But when the objective was placed at one

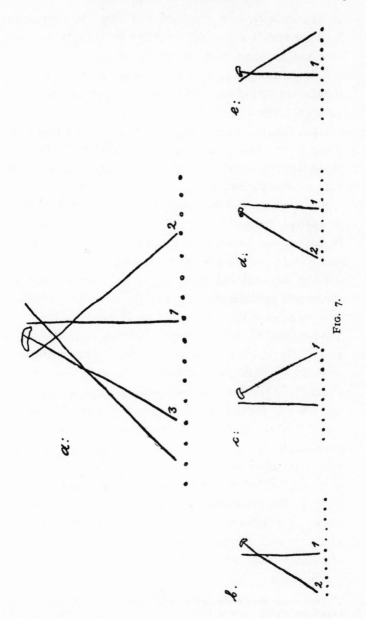

FIG. 7.

metre's distance from the bars, and only two centimetres
from the end of the string, he pulled hesitatingly, but with
eyes and attention fixed on the objective.

A large number of such experiments tend to make the
chimpanzee suspicious. On the whole, however, the results
justify the following conclusions :

When the end of the string or rope is only at a very small
distance from the objective (but much depends on the clearness
of the background) the chimpanzee generally pulls the string,
after a cursory glance at it.

*He will always pull the string if it visibly touches the objective.
It appears doubtful whether the conception of " connexion "
in our practical human sense signifies more for the chimpanzee
than visual contact in a higher or lower degree.*

When rope-end and objective are wide apart, the chimpanzee will generally not pull the rope, unless he is interested
in it *per se*, or wants to have it to use in another way.

When the distance between objective and rope-end is
moderately wide—that is to say, from some centimetres
upwards—so that the rope-end lies in a sort of " halo "
round the objective even to human perception, the result
will entirely depend on the degree of hunger felt by the
animal, and the amount of his attention. When very hungry
the chimpanzee will pull the rope, while looking at the objective, even when he *must and obviously does see* that there
is no contact between them. He then does just the same
thing as the proverbial drowning man, who clutches at a
straw. The movements of the animal in such cases, to
which we shall frequently refer, are slow and give an effect
of complete lack of courage.[1]

---

[1] Hobhouse, *Mind in Evolution* (London 1901, p. 155 seqq.), has made
experiments with ropes on various species of animals ; other scientists
have done the same. We may refer generally to Hobhouse's work,
for other experiments described in the following.

(2)

In the experiments about to be described, the objective is not in any way connected with the animals' room. The only possible implement is a stick, by means of which the objective can be pulled towards the ape.

At the beginning of my work at the station, we had seven chimpanzees. Of these, I found Sultan already quite expert in the use of sticks, and Rana had also been observed performing similar feats. The achievements of some of the other chimpanzees will be recorded later; here we have to do with the three cases of Tschego, Nueva, and Koko.

The full-grown female (Tschego), of whose earlier career in the Cameroons we, of course, know nothing, had been kept almost entirely apart from the other animals up to the time of these experiments (26.2.1914), i.e. one year and six months. She had been in quarters that contained no movable objects, except straw and her blanket, but she was freely permitted to observe the pranks of the young apes. She is let out of her sleeping-place into the barred cage in which she spends her waking hours; outside the cage and beyond the reach of her exceptionally long arms, lies the objective; within the cage, somewhat to one side, but near the bars, are several sticks.

Tschego first tries to reach the fruit with her hand; of course, in vain. She then moves back and lies down; then she makes another attempt, only to give it up again. This goes on for more than half-an-hour. Finally she lies down for good, and takes no further interest in the objective. The sticks might be non-existent as far as she is concerned, although they can hardly escape her attention as they are in her immediate neighbourhood. But now the younger animals, who are disporting themselves outside in the stockade, begin to take notice, and approach the objective gradually. Suddenly Tschego leaps to her feet, seizes a stick, and quite adroitly, pulls the bananas till they are within reach. In this manœuvre, she immediately places the stick on the *farther* side of the bananas. She uses first the left arm, then the right, and frequently changes from one to the other. She does not always hold the stick as a human being would, but sometimes clutches it as she does her food, between the third and fourth fingers, while the thumb is pressed against it, from the other side.

Nueva was tested three days after her arrival (11th March, 1914). She had not yet made the acquaintance of the other animals but remained isolated in a cage. A little stick is introduced into her cage ; she scrapes the ground with it, pushes the banana skins together into a heap, and then carelessly drops the stick at a distance of about three-quarters of a metre from the bars. Ten minutes later, fruit is placed outside the cage beyond her reach. She grasps at it, vainly of course, and then begins the characteristic complaint of the chimpanzee : she thrusts both lips—especially the lower —forward, for a couple of inches, gazes imploringly at the observer, utters whimpering sounds,[1] and finally flings herself on to the ground on her back—a gesture most eloquent of despair, which may be observed on other occasions as well. Thus, between lamentations and entreaties, some time passes, until—about seven minutes after the fruit has been exhibited to her—she suddenly casts a look at the stick, ceases her moaning, seizes the stick, stretches it out of the cage, and succeeds, though somewhat clumsily, in drawing the bananas within arm's length. Moreover, Nueva at once puts the end of her stick behind and beyond the objective, holding it in this test, as in later experiments, in her left hand by preference. The test is repeated after an hour's interval ; on this second occasion, the animal has recourse to the stick

---

[1] As is well known the chimpanzee never sheds tears.

much sooner, and uses it with more skill; and, at a third repetition, the stick is used immediately, as on all subsequent occasions. Nueva's skill in using it was fully developed after very few repetitions.

On the second day after his arrival (10.7.1914), Koko was, as usual, fastened to a tree with a collar and chain. A thin stick was secretly pushed within his reach; he did not notice it at first, then he gnawed at it for a minute. When an hour had elapsed, a banana was laid upon the ground, outside the circle of which his chain formed a radius, and beyond his reach. After some useless attempts to grasp it with his hand, Koko suddenly seized the stick, which lay about one metre behind him, gazed at his objective, then again let fall the stick. He then made vigorous efforts to grasp the objective with his foot, which could reach farther than his hand, owing to the chain being attached to his neck, and then gave up this method of approach. Then he suddenly took the stick again, and drew the objective towards himself, though very clumsily.

On repeating this experiment I was even more struck by the clumsiness of this animal; he often pushed the banana from the wrong (hither) side, so that it was once sent to quite a distance from him. In this case—and frequently on other occasions—Koko used his foot to grasp the stick, and continued to make vain efforts in this manner. Finding it of no avail, he suddenly took up a green stalk, with which he had been playing before the experiment began, but this was quite useless, the stalk being even shorter than the stick. From the beginning, Koko held the stick in his right hand, and only had recourse to the left for a few minutes when his feeble muscles were obviously tired; but when using his left hand, the stick wobbled aimlessly from the beginning (it could not have been from exhaustion), and was immediately transferred to the right.

It may be accepted as a general axiom that a chimpanzee who has once begun to use a stick for these purposes is not quite helpless if there is no stick to hand, or if he does not perceive one that is available.

Two days later, after she had played with it a good deal, Nueva, for the following test, was deprived of her stick (13.3). When the objective was put down outside the cage, she at once tried to pull it towards her with rags lying in her cage, with straws, and finally with her tin drinking-bowl which stood in front of the bars, or to beat it towards her—using the rags—and sometimes successfully.

On the day after Tschego's first test, two sticks lay inside the cage, about one and a half metres from the bars. When Tschego was let into her cage, she at first stretched her arm out through the grating towards the fruit; then, as the youngsters approached the coveted prize, Tschego caught up some lengths of straw, and angled fruitlessly with them. Only after a considerable time, as the young apes approached dangerously near to the objective, Tschego had recourse to the sticks, and succeeded in securing it with one of them.

In the next test, which took place several hours later on the same day, the sticks were removed to a greater distance from the bars (and, therefore, from the objective beyond them) and placed against the opposite wall of the cage, four metres from the grating. They were not used. After useless efforts to reach the bananas with her arm, Tschego jumped up, went quickly into her sleeping-den, which opens into the cage, and returned at once with her blanket. She pushed the blanket between the bars, flapped at the fruit with it, and thus beat it towards her. When one of the bananas rolled on to the tip of the blanket, her procedure was instantly altered, and the blanket with the banana was drawn very gently towards the bars. But the blanket is, at best, a troublesome implement; the next banana could not be caught like the first. Tschego looked blank, glanced towards the sticks, but showed not the least interest in them. Another stick was now thrust through the bars, diagonally opposite to the objective, Tschego took, and used, it at once.

Koko, who had already tried to use a plant-stalk in the same circumstances, three days later (13.7) in the course of the test, ignored the stick which lay a little to one side and on the periphery of his " sphere of action." Only after some time did he grasp the stick with his foot, and thus drew the bananas, clumsily enough, towards him. On a repetition of the experiment, he fetched his blanket and dragged it close to the objective, then let it fall after a short hesitation, and took up the stick once more. A day later, when no stick was available, he repeated the blanket procedure exactly, and then tried to angle the objective with a stone. Some days after he employed a large piece of stiff cardboard, a rose-branch, the brim of an old straw hat, and a piece of wire. All objects, especially of a long or oval shape, such as appear to be movable, become " sticks " in the purely functional sense of " grasping-tool " in these circumstances and tend in Koko's hands to wander to the critical spot.

[Incidentally, an observation on myself: Even before the chimpanzee has happened on the use of sticks, etc., one expects him to do so. When he is occupied energetically, but, so far, without success, in overcoming the critical distance, anxiety causes one's view of the field of action to suffer a phenomenological change. Long-shaped and moveable objects are no longer beheld with strict and static impartiality, but always with a " vector " or a " drive " towards the critical point.]

As is to be expected, variations in the nature or position of the objective have very little influence on the use of the stick, when that instrument has once been mastered. One hot day Koko even tried to pull a pail of water, which had been left standing in his neighbourhood, towards him by a stick held in each hand; of course, without succeeding. When the bananas are hung out of reach on the smooth wall of the house, he takes a green plant-stalk, then a stone, a stick, a straw, his drinking-bowl, and finally a stolen shoe, and stretches up towards the fruit; if he has nothing else to hand, he takes a loop of the rope to which he is attached, and flaps it at the bananas.

When animals who have developed behaviour to cope with the requirements of a special given situation use the same methods in situations, only similar or partially similar, their observers conclude, often correctly no doubt, that the cloudy perception of the animal sees no difference between the two situations, and, therefore, adopts the same procedure in each. It would be a mistake to give such an explanation when the chimpanzee replaces his stick by other objects. The vision of the chimpanzee is far too highly developed —as can easily be proved both by tests and by general observation—for him to " confuse " a handful of straw, the brim of a hat, a stone, or a shoe, with the already familiar stick. But if we assert that the stick has now acquired a certain *functional or instrumental value* in relation to the field of action under certain conditions, and that this value is extended to all other objects that resemble the stick, however remotely, in *outline* and *consistency*—whatever their other qualities may be—then we have formed the only assumption that will account for the observed and recorded behaviour of these animals. Hats and shoes are certainly not visually identical with the stick and, therefore, interchangeable in the course of the test experiments; *only in certain circum-*

stances are they functionally sticks, after the function has
once been invested in an object which resembles them in
shape and consistency, namely a stick. As has been shown
in the account of 'Koko's behaviour, practically no limitation
with regard to type remains in the case of this youngster,
and almost every " movable object " becomes, in certain
circumstances, a " stick."

A far more important factor than the external resemblances
or differences between stick, hat brim, and shoe, is in the
case of Tschego and Koko the *location of the implement* both
in relation to the animals themselves and the objective.
(Nueva was not tested in this manner, for some reason.)
Even sticks that have already been used often both by Tschego
and Koko seem to lose all their functional or instrumental
value, if they are at some distance from the critical point.
More precisely : if the experimenter takes care that the stick
is not visible to the animal when gazing directly at the objective
—and that, vice versa, a direct look at the stick excludes
the whole region of the objective from the field of vision
—then, generally speaking, recourse to this instrument is
either prevented or, at least, greatly retarded, even when it
has already been frequently used. I have used every means
at my disposal to attract Tschego's attention to the sticks
in the background of her cage (see above) and she did look
straight at them ; but, in doing so, she turned her back on
the objective, and so the sticks remained meaningless to
her. Even when we had induced her, in the course of one
morning's test, to seize and use one of the sticks, she was
again quite at a loss in the afternoon, although the sticks
had not been removed from their former position, and she
stepped on them in the course of her movements to and
fro, and repeatedly looked straight at them. At the
same time, *sticks*—and other substitutes—*which she beheld
in the direction of her objective, were made use of without*

any hesitation, and she devoured what food she could reach
with relish.[1]

We subjected Koko to a similar test with similar results.
He made useless efforts to reach the objective : a stick was
quietly placed behind him ; but though, on turning round,
he looked straight at the stick and walked across it, he did
not behold in it a possible implement. If the stick was
silently moved towards him, so that the slightest movement
of head or eyes would lead from the region of the objective
to the stick—suddenly he would fix his gaze on it, and use it.[2]

The important factor here is not only the distance of the
stick from its objective ; for instance, suppose that Koko is
seated in the centre of his chain-circle, the objective set
down outside the circumference, and midway between ape
and objective is placed a stick : Koko will then generally
pick up the stick on the way to the objective, and naturally
so ; for his glance towards the goal can hardly miss the stick
in this case, and it is highly probable that he *sees them " in
connection,"* which would be favourable to the result.

There is no absolute rule here, however. It sometimes
happens that some useful object, at quite a distance behind,
is noticed as the animal looks back, and fetched. Such a
result is only to be expected in the variety of circumstances
that are at work ; but as a rule, and to a conspicuous degree,
the behaviour was as described in the preceding pages.

[1] The blanket (see above) lay in the animals' sleeping den, as far behind
as the sticks, and nevertheless was fetched : yes, but the open door of
the den is close to the bars, in the foreground of the cage, so that Tschego,
with a slight movement of the head, which permits the bars and the
objective to remain in her field of vision, can also see the blanket.
Whereas, if she faces the sticks, the whole region of the objective
vanishes. Besides, the blanket is seen and used daily, and is thus
sui generis and in a different category to other objects.

[2] The animal must not see the stick in motion, i.e. when it is being
pushed towards him ; that would bring quite new conditions into play.
I removed Koko or held my hand over his eyes, when I moved the
stick ; in the first case I replaced him before the objective in his exact
former position. Such young animals can be handled easily, and Koko
was quite used to it.

We found, however, that, although to some degree the use of the stick as an implement depends on the geometrical configuration, this is only so on first acquaintance. Later on, after the animal has experienced frequently the same conditions, it will not be easy to hinder the solution by a wide optical distance between objective and implement. But one can oneself " feel " that at the *inception* of these tests there is a dependence (on spatial position) such as has been described above. If one asks where to place the stick, the conviction arises—at once and without any previous reasoning—that the solution will be specially easy, if the stick is in the immediate neighbourhood of the objective, and can be visualized in connexion with the objective. However familiar the procedure in this situation may have become to us, we still dimly apprehend the decisive factors.

(3)

When the objective is fastened *at a height* from the ground, and unobtainable by any circuitous routes, the distance can be cancelled by means of a raised platform or box or steps which can be mounted by the animals. All sticks should be removed before this test is undertaken, if their use is already familiar—the possibility of utilizing old methods generally inhibits the development of new ones. (24-I-1914). The six young animals of the station colony were enclosed in a room with perfectly smooth walls, whose roof—about two metres in height—they could not reach. A wooden box (dimensions fifty centimetres by forty by thirty), open on one side, was standing about in the middle of the room, the one open side vertical, and in plain sight. The objective was nailed to the roof in a corner, about two and a half metres distant from the box. All six apes vainly endeavoured to reach the fruit by leaping up from the ground. Sultan soon relinquished this attempt, paced restlessly up and down, suddenly stood still in front of the box, seized it, tipped it hastily straight towards the objective, but began to climb upon it at a (horizontal) distance of half a metre, and springing upwards with all his force, tore down the banana. About five minutes had elapsed since the fastening of the fruit ; from the momentary pause before the box to the first bite into the banana, only a few seconds elapsed, a perfectly continuous action after the first hesitation. Up to that instant none of the animals had taken any notice of the box ; they were all far too intent on the objective ; none of the other five took any part in carrying the box ; Sultan performed this feat single-handed in a few seconds. The observer watched this experiment through the grating from the outside of the cage.[1]

There was something clumsy in the animal's execution of this feat. He could quite well have shoved the box right under the bananas ; as he trundled it along just before his jump, the open side of the box came uppermost ; Sultan did not alter this, but stepped on to the edge of the box, which was, of course, less convenient as a " take-off." Also, he did not place the box lengthways vertically which would have saved waste of energy. But the whole action was too rapid to allow of delay for these refinements.

On the following day the test was repeated, the box being placed as far from the objective as the available space permitted, i.e. at a distance of five metres. As soon as Sultan had grasped the situation, he took the box, pulled it along till it was almost directly beneath the bananas, and jumped. On this occasion a covered side of the box was uppermost.

[1] Except in a few cases, which will be fully recounted, the observer's rôle is only that of preventing the easiest methods of procedure, detours in the ordinary sense. He can be present without damaging the test ; the chimpanzees take little notice of him. It is, of course, understood that he is absolutely neutral and passive, except when the contrary is explicitly stated.

We shall describe elsewhere the results of this box-test in the cases of Tschego and the other five animals of this group. Nueva unfortunately died before she could be tested. Koko was faced with this problem, and gave most curious results.

On the third day of his residence at the station (11.7), he was given a small wooden box as a toy. (Its dimensions were forty by thirty by thirty centimetres.) He pushed it about and sat on it for a moment. On being left alone, he became very angry, and thrust the box to one side. After an hour had elapsed, Koko was removed to another place. There his chain was fastened to the wall of a house. On one side, one metre from the ground, the objective was suspended from the wall. The box had been placed between three and four metres from the objective, and two metres from the wall, while Koko was being conducted to his new place. The length of his rope allowed him to move freely about the box and beside the wall where the objective hung. The observer withdrew to a considerable distance (more than six metres from the box, and the same side), and only approached once in order to make the objective more attractive. Koko took no notice of him throughout the course of the test. He first jumped straight upwards several times toward the objective, then took his rope in his hand, and tried to lasso the prize with a loop of it, could not reach so far, and then turned away from the wall, after a variety of such attempts, but without noticing the box. He appeared to have given up his efforts, but always returned to them from time to time. After some time, on turning away from the wall, his eye fell on the box : he approached it, *looked straight towards the objective*, and gave the box a slight push, which did not, however, move it ; his movements had grown much slower ; he left the box, took a few paces away from it, but at once returned, and pushed it again and *again with his eyes on the objective*, but quite gently, and not as though he really intended to alter its position. He turned away again, turned back at once, and gave the box a third tentative shove, after which he again moved slowly about. The box had now been moved ten centimetres in the direction of the fruit. The objective was rendered more tempting by the addition of a piece of orange (the *non plus ultra* of delight !), and in a few seconds Koko was once more at the box, seized it, dragged it in one movement up to a point almost directly beneath the objective (that is, he moved it a distance of at lease three metres), mounted it, and tore down the fruit. A bare quarter of an hour had elapsed since the beginning of the test. Of course, the observer had not interfered with either the ape or the box, when he " improved " the bait. The enhancement of the prize by the addition of further items is a method which can be employed over and over again with success when the animal is obviously quite near to a solution, but, in the case of a lengthy experiment, there is the risk that fatigue will intervene and spoil the result. It must not be supposed that before the exhibition of the orange, the animal was too lazy to attain its objective ; on the contrary, from the beginning, Koko showed a lively interest in the fruit, but none—at first—in the box, and when he began to move it, he did not appear *apathetic* but *uncertain :* there is only one (colloquial) expression that really fits his behaviour at that juncture : " it's beginning to dawn on him ! "

There are other definite proofs that the animal did not hesitate out of pure laziness to employ a method that he would have understood quite well. For instance : the test was repeated after an interval of a few minutes, the objective this time was fastened to the same wall, but on the other side of the fastening which held Koko's rope to the wall, and more than three metres from its former position ; the box was left untouched, standing just under the former site of the objective, where Koko had himself dragged it. He sprang at the new objective in the same manner as before, but with somewhat less eagerness ; at first he ignored the box. After a time he suddenly approached it, seized and dragged it the greater part of the distance towards the new objective, but at a distance of a quarter of a metre he stopped, gazed at the objective, and stood as if quite puzzled and confused. And now began a tale of woe for both Koko and the box. When he again set himself in motion, it was with every sign of rage, as he knocked the box this way and that, but came no nearer to the objective. After waiting a little the experiment was broken off, so that the box should not be knocked unintentionally in the direction of the fruit, and thus afford a solution by pure chance.

The next day the box was again ignored although Koko was much exercised to reach the objective, and tried the most varied implements, among others the aforementioned old shoe, in lieu of a stick. Occasionally he laid hold of the box, but it was not evident that he had the objective in view as he did so. Two days later the scene was changed, the fruit being fastened to another wall, and the box placed four metres away from it. The animal used every means —except the box—but could not reach the prize. After one of these vain attempts, on turning away, the box caught his eye, and he looked fixedly at it, so that the observer momentarily expected him to fetch it. But Koko looked away again, and then tried a new method which is described below. As that, too, was unsuccessful, he sat down, exhausted, on the box and presently began to hop about playfully, still seated on the box. In the course of the next test, two days later (16.7), it became increasingly evident that the solution had gone from his recollection. In this experiment, two boxes were placed about five metres from the objective. Koko eyed them suspiciously from time to time, but did not fetch them, keeping to other methods of procedure. Finally, covering Koko's eyes with our hands, we placed one of the boxes so close to the wall that—as he forthwith demonstrated—he had only to stand on it to touch the wall immediately under the objective with his hand. Therefore, if the box were pushed a very little further, the thing was done. Koko stretched and strained himself as he stood on the box, but he did not give the slight propulsion that was alone necessary. The next day he was allowed to play with the box for a short time ; among his movements we noticed the following : throwing over the box, hopping on it, and sitting in it. Five days later (21.7), on the occasion of the next experiment, Koko used anything and everything he could lay hand on as a substitute for the familiar stick ; and the box he merely stared at frequently in a peculiar manner. Suddenly he flew at it and began a violent attack : he was beside himself with rage, flung the box to and fro, and kicked it. These outbreaks which had been rarer on previous days and were considered the result of accumulated ill-humour, were now concentrated entirely on the box. Again and again as he turned from the objective, his eyes sought the box ; he glared, and then fell upon it.

After an interval of nine days, the experiment was continued (30.7). In the intervening period Koko was not shown the box at all. The objective was hung on the wall as before, and the box placed two metres away from it (diagonally). Koko stretched towards the prize for a while in vain ; he turned away, saw the box, and stared at it for a moment. He went up to it and seized it ; for a minute it looked as though he was about to attack it again ; but instead, he dragged it hurriedly beneath the objective, mounted it and pulled down the fruit. The place in which this test was completed was not the scene of the first experiment, and between the two dates there was an interval of nineteen days, during the first half of which no trace of a solution appeared, except an equivalent of the words : " there's something about that box."

After this Koko did not lose mastery of the solution, although in the two repetitions of the test which immediately followed his success, he at first stretched up and leapt towards the fruit ; but the box was presently brought up after all. In the second repetition Koko had set down the box too far from the objective in his haste, so could not reach the prize when standing on it. He got down at once and pushed the implement into the right position. The initial distance between objective and box did not seem to exert any influence, the box being carried for six metres as quickly as for two. The following day Koko began to turn towards the box and seize it as soon as anyone came in sight carrying edibles. He often picked up a stick, too, and carried it with him ; he either threw the stick away as he mounted the box, or on other occasions hooked the fruit off the nail with it. Such a mixture of methods was observed in other chimpanzees also : sometimes they were ingenious and, in fact, the only practicable means to the desired end.

[I have thought proper to give this account in minute detail, as the behaviour of the animal showed such variations—perhaps on account of his extreme youth—and from a theoretical point of view was so much more suggestive than if every test had gone off smoothly. One can only try to explain these achievements by observing them in detail. It is also, in my opinion, quite as needful to the understanding of the chimpanzee to discover what behaviour led him to his solutions, as to know that he " uses a box as a tool " at all.]

Variations of the Test.

The day after Sultan had used the box for the second time, the objective was fastened to the roof of another room which was at a much greater height from the ground. Two boxes stood close together on the ground, about five metres from the fruit. Sultan was alone. At first he took no notice of the boxes, but tried to knock down the objective, first with a short stick and then with one of more appropriate length. The heavy sticks wobbled helplessly in his grasp he became angry, kicked and drummed against the wall and hurled the sticks from him. Then he sat down on a table, in the neighbourhood of the boxes, with an air of fatigue; when he had recovered a little, he gazed about him and scratched his head. He caught sight of the boxes—stared at them, and in the same instant was off the table, and had seized the nearer one of them, which he dragged under the objective and climbed upon, having first re-captured his stick, with which he easily secured the prize. The box was not placed vertically, and Sultan was too inexpert at jumping to be able to dispense with his stick.

The next day we removed the sticks, but placed boxes and objective as before. A light table, which was not used in the former test, stood in the same place, about three metres from the objective. Sultan made many fruitless attempts. He pulled one of the boxes beneath the prize, but after obviously measuring[1] the distance with his eyes, he did *not* mount the box, which would have been useless in any case, but pushed it hesitatingly to and fro beneath the fruit. One corner of the box happened to land on a thick beam which lay on the ground a little to one side. Sultan gazed upwards, but the distance was too great even then, and he fell upon the box in a fit of anger. Presently he took notice of the second

[1] The term " measuring " is no " anthropomorphism ". At any time it may be observed that a chimpanzee, before making a wide jump at a considerable height, looks carefully to and fro across the intervening space. As an arboreal animal with immense range of spring and the need to use it, he *must* be able to measure distances. It would be quite unjustifiable to object to the use of the term " measuring " in this connexion.

box and fetched it, but, instead of placing it on top of the first, as might seem obvious, began to gesticulate with it in a strange, confused, and apparently quite inexplicable manner ; he put it beside the first, then in the air diagonally above, and so forth. This state of unordered confusion was followed by the customary paroxysm of anger ; he seized the intractable box and rushed up and down the room, bumping the box behind him and dashing it with his whole strength against the wall. When his rage had spent itself, he gave a calm, quiet look at the scene before him and made a long step in advance by *lifting* the first box, which was still directly beneath the objective, and placing it upright on end with a powerful and dexterous movement. Unfortunately a second look showed him only too clearly that even thus he could not reach the objective, and he did not mount the box. He turned instead towards the beam against which the box had been wedged and, by dint of his utmost exertions, lifted it at the end nearest to the objective, but could not raise it high enough for his purpose. After his second disappointment, he again gazed round helplessly and finally noticed the table.[1] He seized it by one leg and dragged it towards his goal, but turned it over through his hasty, jerky movements. Had he brought it under the objective, his problem would have been solved. As the only resemblance between table and box is that both are made of unpainted wood, this must be either an example of the discovery of a new method, or a substitution to which our remarks on the case of stick-substitutes are entirely applicable. It is absolutely impossible that Sultan should simply " confuse " box and table.

Immediately after the test described above, the following experiment was undertaken : the table was removed and a small ladder of five rungs, one metre thirty centimetres

---

[1] This was not the same table on which he had sat down to rest himself the previous day, and which apparently was too heavy and firmly fixed into its corner to become a tool.

in length, was put down, not in the same place, but at about
the same distance[1] from the objective, which was hanging
from the roof near a wall. After a few seconds, Sultan
took up the ladder, pulled it underneath the objective and
tried hard to lift it into position. As the result of a most
peculiar proceeding which is described below, he only succeeded in mounting the ladder and securing the prize after a
considerable interval.

When the other apes had familiarized themselves with
the use of the box, their behaviour showed no appreciable
difference from Sultan's. Therefore it is quite permissible
in everything that follows to use also the observations made
on them ; they were strongly aided by others at the first use
of the procedure, but they varied the method quite independently later on, and in quite a similar manner as Sultan.

We have watched all of them gradually replacing box,
ladder and table, by a highly miscellaneous collection of
objects ; stones, iron grills from the windows of the cages,
tins, blocks of wood, coils of wire, all these were indiscrimin-
ately collected and employed as ladders or footstools—
objects which in the practice of chimpanzees are almost
identical functionally. But the most curious variation is
that recounted below, adopted by Sultan immediately after
his first attempt with the ladder, when he could not stand
it in position. In order to urge the now fatigued chimpanzee
towards fresh efforts, the observer emerged from his usual
place in the background, and approached within an arm's
length of the objective, to which he pointed. Suddenly
Sultan jumped up, seized the observer's hand, and tried with
all his strength to drag him towards the fruit. As I received
the impression that Sultan wanted to make me *give* him
the fruit, I shook him off, and then, as he continued with
the greatest persistence to seize and drag at my hands and

[1] The distance was about three metres.

feet, I pushed him abruptly away. He fell into a paroxysm of fury, with the accompanying symptoms of throat convulsions and erection. Presently the keeper passed by crossing the room beneath the objective ; Sultan walked quickly up to him, took his hand, pulled him in the direction of the fruit, which was behind him, and made unmistakable efforts to climb onto his shoulder. The keeper freed himself and moved as far away as the space allowed, but Sultan would not let him be and pulled him back again till he was underneath the goal. At my instructions the keeper made only a pretended resistance, and it only took a few seconds before Sultan had sprung on his shoulder, and torn down his prize. The animal was now absolutely " set on " this easy method of solution, but before he was, in the interest of the experiments, taught to relinquish it, there took place several violent scenes, during which Sultan more than once appeared on the point of suffocation.

A further modification of this method, namely, the employment of another chimpanzee as a footstool, was spontaneously introduced by little Konsul, who, *nota bene*, like the other apes, had not seen Sultan using us as ladders. The circumstances were particularly happy : Konsul was in the habit of walking directly behind one of the other animals and, so to speak, " in his footsteps ", placing both hands on his elder's shoulders and keeping step with him. As a rule, the larger animal made no objection ; on the contrary, one of the others would often place Konsul's hands on his own shoulders, as an invitation so to accompany him.[1] In the course of one of the

---

[1] Tschego showed for months a strong friendship for the little creature. When she emerged from her den and joined the others, Konsul attached himself to her in the manner described above, or (later on) sprang on to her shoulders and let her carry him like a horse. I do not know whether the ape-mothers sometimes carry their children thus. In this case the " keeping step " of Konsul's would be a kind of " survival ". (*Suckling* infants they carry in front of the lower abdomen ; see von Allesch, *Berichte der Preuss. Akad. d. Wissenschaften*, 1921.) The only other animal I observed " keeping step " was Chica—and she did so rarely.

tests, in which the objective was hung from the roof, Konsul was thus walking up and down, supporting himself on Grande's shoulders, when, during their walk, they once approached close to the objective; Konsul made a hasty effort to climb onto Grande's back, and succeeded in doing so but only after Grande, who had no idea of what was intended, had passed the point of vantage. As the pair passed again beneath the fruit, the same incident was repeated and the result was, that *Sultan* now successively clutched observer, keeper and then (when we pushed him away) first Tercera and then Rana and tried to drag them under the objective. Both Rana and Tercera fled before him in dismay; they could not understand why he pursued them with outstretched hand, and feared an attack. Finally he succeeded in pinning down Rana and springing onto her shoulder; but as she crouched close to the ground in terror, he had to make several jumps before he succeeded in grasping the prize, and each time he fell back heavily on the prostrate Rana. From this time onward similar occurrences were frequent; for instance, on the following day, Konsul tried to climb onto Grande, Sultan successively got on to Rana, Grande, and Tercera, and finally Rana on to all the rest together, for there was now a struggling group of chimpanzees, who all gripped each other, and lifted their feet to climb, but none of whom wanted to be footstool. In the course of the same experiment I hung up the fruit in the presence of the animals; as I stepped back, I was seized from behind and held fast; it was Grande, who climbed quickly on to my shoulder and reached the prize. I put up a new goal and retreated as rapidly as possible, but Grande quickly followed me with outstretched hand and lips distended, uttering whimpering sounds, and dragged me into position. The animal was already well acquainted with the use of boxes and was, therefore, perhaps, especially quick to grasp the similar functional value of a human body.

In the same manner as Grande and Sultan, Chica and Rana later on used me, the keeper, or anyone else they could get hold of, as a footstool.

The technique of the use of boxes was forthwith applied to somewhat different siuations. Sultan, for instance, was pursuing one of the others, who was fleeing along the wire-roofing ; he was not a sufficiently expert climber to pursue his fellow along the roof, but seized a box, put it beneath the other ape, and sprang from it. The box was too low so he tried to fetch Grande—and so forth. Later, when they all knew the use of boxes and other implements, they unfortunately became used to dragging their tools under such positions of the roof as were lower than the rest, and thus accessible from a relatively slight height.[1]

The method of climbing upon each other was first an interlude in a test, during which the implement was *not* directly visible, but could only be drawn into the situation in consequence of "remembering." (February 15th.) The room from whose roof the objective was suspended communicated by a door with the corridor which, at a distance of eight metres from the door, turned at a right angle. Round the turn of the corridor stood the ladder, which was therefore completely invisible from the room containing the fruit. *Before* the test, the apes were allowed to play freely in the passage, where they could see the ladder but not the room, for the door was still shut. Sultan showed that he, at least, did notice the ladder, for he gnawed persistently at one of the uprights. After the door was once opened, the objective exercised such a powerful attraction that not one of the animals stayed in or returned to the passage. One effort after another (see above) was made to reach the prize, but even Sultan did not remember the ladder. Finally he was

[1] As the wire was thin, it was often possible at this time to observe the chimpanzees in a state of complete freedom.

led by the hand into the corridor and past the ladder, but
without having his attention specially called to it. There
was no immediate effect; on his return to the room, he tried
as before to use the others as footstools. Immediately
there was a violent set-to among the animals, so that the
observer was forced to intervene and separate them; when
peace had been restored, Sultan was missing. A bumping
sound proceeded from the corridor, and he reappeared dragging
the ladder behind him. As in this instance my observation
of the test had been interrupted and the crucial moment had
escaped me, I repeated it on the following day, but used a
box, placing it exactly where the ladder had stood; I took
care that Sultan should have seen it before the test began
and fastened up the fruit as on the former occasion. Sultan
made one effort after the other towards solving his difficulty;
e.g. he pulled one of the iron bars in Tschego's sleeping-den[1]
out of its socket, propped it like a ladder against the wall
and swarmed up it towards the fruit. But all his efforts
were made in the neighbourhood of the objective; he appeared
to have forgotten the box. After waiting a long time, I
took Sultan's hand, led him up to the box, past it—without
drawing his attention to it—and back; but the only immediate result was that he clutched my hand more tightly as
we returned, and tried to pull me under the objective. When
this failed, he took the greatest trouble to find something in
the immediate vicinity that he could use as a tool, and had
recourse to a long bolt which was fastened to the outer side
of the door; he hung on to the door, which was half open,
and tore at the bolt with all his power. This time the observation succeeded: *quite abruptly, and without visible external
cause*, Sultan ceased belabouring door and bolt, *remained
for a moment motionless*, sprang to the ground, traversed the

---

[1] This den was the *first* room opening from the corridor, and just next
to the open door.

passage at a gallop, and was back in a moment with the box.
In that second, in which his behaviour obviously took a
fresh direction, the door covered and concealed the objective
from his view, which did not prevent him from trying to tear
away the bolt as an implement : yet the box was at a much
more considerable distance, round the corridor corner, and
behind his back. It is evident, however, how immensely
delayed the solution may become when the adequate implement can be introduced only through the action of memory.
Sultan had already passed through the same test (with the
ladder) the day before ; nevertheless, it was possible *during
the test*, when there must have been a strong drive towards
the objective, to lead him past the familiar implement *without
the solution suggesting itself to him* ; but of course this excursion
of a few seconds' duration certainly brought him out of
" the region of the objective ". At the end of the passage
even the bolt appealed to him as an implement, although
the objective was not visible to him simultaneously ; but
the whole region of the door was in immediate contact with
the room in which the experiments took place. The difficulty
here therefore seems similar in kind to—though much greater
in degree than—those which had faced Tschego and Koko,
owing to the unusual positions in which their sticks were
placed. The best tool easily loses its situational value if it
is not visible simultaneously or quasi-simultaneously[1] with
the region of the objective.

In the test now to be described the problem and its solution
are externally different from the former examples, but in
principle they are similar. The apes had all become expert
in the use of boxes by the time this experiment took place.

D1 to D4 represent the four doors of the Ape-house H,
opening on to the playground P ; the doors are identical in
appearance and situated at equal intervals of distance from

[1] This term hardly requires explanation.

one another; O is the objective which hangs from the wire roof, but so high that it cannot be reached from the ground. The precise point from which it was suspended was chosen so as to be directly opposite to the hinges of the door D2, at a distance slightly exceeding the width of the door. It was at about the same height from the ground as the lintel of the door. The apes had now got out of the habit of climbing along the wire-roof; they no longer ventured to do so during the tests, at any rate.

(April 12th.) The doors 1, 3 and 4 had been shut; door 2 was not latched, but very careful attention was necessary to perceive that it was "different" from the other three.

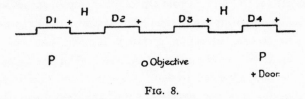

FIG. 8.

Sultan was then fetched. On perceiving the objective, he picked up a short stick which he forthwith threw away without attempting to use it—it was much too small. Immediately after, his glance fell on D2, at which he stared fixedly for several seconds, without moving. Finally he went up to it, opened it—still standing on the ground—and climbed upon it. As he had not opened the door at a full right angle to the wall, he could not quite reach the objective by this means. He therefore dismounted, pulled it wide open and then climbed on to it again; he would have reached the objective if his weight had not swung the door somewhat back from the right point. So he stopped, again descended, and, placing the door in the correct position, captured the prize without further trouble. His correction of the procedure in the beginning of this experiment and the "compensation" of the door's swing back, were carried out with a lucidity which no human being could have surpassed, very much in contrast to his behaviour in certain other situations, as will be recounted later.

As Sultan had succeeded in this experiment so rapidly, it was repeated with Rana, who was unquestionably the least intelligent of our animals (April 14th). She entered the room, beheld the objective, and gazed towards the door. Then she climbed along the beams to the lintel and pushed the door off from the wall, so that she was able to secure the objective, seated on the top of the door and travelling with it.

The door which both animals utilized is that of the room in which Sultan spends his nights. Rana sleeps in the room behind D1. Both have often perched on the doors : at

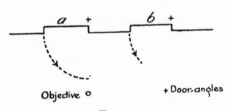

Objective o   + Door-angles

FIG. 9.

one time Sultan as he ate his meals used to squat on door D2, which was then thrown quite back and hooked to the wall. Undoubtedly they had both had occasion to propel doors to and fro, when in this position ; but what is novel in this situation is the presence of the objective at right angles to the hinges. Their previous acquaintance with the doors, it is true, greatly facilitated their success in this test.

(Test on May 6th.) All four doors were pushed to in the same manner, without being latched. The objective was suspended in front of D3 instead of D2. Rana was tested alone on this occasion, as before. She grasped a stick and climbed up the wall with it in her hand, almost opposite the objective, lifted the stick once, dropped it, pushed door D3 open, and thus attained the fruit.

In the course of similar variations on this test it may happen that one of the animals does not open the most convenient door, but the one adjacent; e.g. (cf. Fig. 9) door *b* instead of door *a*. But this action, too, is to some extent directed towards the objective, and the door is actually only opened so far as to be at its nearest to the prize. Possibly the fact that only a slight propulsion of door *b* brings it directly opposite their goal will account for this. We shall see—in the second series of our investigations—that the turning of the doors and the form of space traversed thereby are less comprehensible to the apes than simpler movements and their corresponding forms of space.

[Tschego's behaviour in this experiment was quite clear. This adult ape was, as a rule, too lazy and too heavy to be used in all the tests suitable for the youngsters. Generally she was present during their performance and appeared to take no notice of them. She had seen the manipulations of the door, in the course of the experiment just described squatting apparently passively in Rana's immediate neighbourhood. A new objective was put in position, and all the doors were closed but not latched. After an interval of apparent total indifference, Tschego slowly got on her feet, approached the right door, opened it at right angles so that it pointed straight at the fruit, and laboriously climbed up the *outer* side of the door—that facing the playground. It was the first time she had done so and she was truly no expert at gymnastics ! Under its heavy load the door swung slowly back. Tschego at once ceased her scramble up the door, descended, opened the door again at right angles and ascended once more, with the same result. She again returned to the ground, opened the door carefully at right angles and climbed it *edgeways* with infinite trouble ; but always the door received an impetus, though but a feeble one, and began to move slowly away from the desired goal. After having once more corrected the position, Tschego climbed again, with unusual celerity and from the *inner (room) side :* the door remained stationary and she attained the fruit. It is probable, of course, that the behaviour of the young apes of which she had been witness influenced Tschego to some extent ; but the triumphant final tactic was entirely her own achievement ; only on one previous occasion (Sultan's first test) had the door swung on its hinges away from the objective, and then he did not change the side which he climbed.]

(4)

A gymnastic bar about two and a half metres high had suspended from its projecting end a strong rope, with which the animals often played. The objective had been hung from the roof on a level with the top of the frame, two metres from the rope and about two from the ground.

(February 27th.) Sultan tried to lift or drag along the heavy ladder which had been used to bring the objective into position, and was lying close by ; then he turned his attention to a heavy board—also in vain. After having once more turned to the ladder, he climbed the gymnastic bar, and saw from that eminence a broken broom : he descended, picked up the broom, and returned to his point of vantage, where he endeavoured to knock down the fruit with it. He could not succeed, of course ; so still armed with the broken broom, he returned to the ground, and tried to use the miserable implement as a "jumping-stick" (cf. below) from beneath the goal, but almost at once ceased the hopeless effort. He pulled about the heavy board and the ladder once more. Then he tried to use the observer as a substitute for the ladder. On being repulsed, he returned to the gym-nastc bar. He seized the rope and swung towards the objective, but with little energy, as if it were a hopeless undertaking ; he then climbed upon the upper bar and squatted there, staring fixedly at the fruit, and with an attitude and expression which in a human being anyone would have described as "thoughtful." As his hesitation in handling the rope had suggested that he was afraid to swing as far as was necessary—for Sultan was not only a mediocre gymnast, but also in his childhood quite sufficiently cautious, to say the least—the objective was hung lower and a little nearer to him. A few seconds later Sultan seized the rope, swung himself up with sufficient energy and tore down the fruit. Neither during the first half of the experiment, nor while the place of the objective was being altered, nor afterwards, was his attention drawn by the observer to the rope.[1]

Sultan was removed, the fresh fruit hung up in the *latter* position of the old objective, and Chica, accompanied by Tercera, was let into the room. After the two had recovered from the dread of being alone, they began to take an interest in the objective. After gazing up at it, Chica mounted the bars, pulling the rope with her. She squatted on the frame, swung the rope-end out towards the fruit, as though to knock or lasso it down, but the distance was too great for her. She presently relinquished this method ; climbed down part of the way, still holding the rope, swung herself vehemently forward, and captured the prize.

(March 7th.) Grande and Rana, in the same circum-stances, simultaneously approached the bars and at once and simultaneously seized the rope and tried to swing them-selves towards their goal. But this effort failed—for physical

[1] In future I must excuse myself from further repetition of this statement. *Every* indication and *every* assistance given are expressly mentioned in the accounts of the tests.

reasons—and Rana retreated before the formidable Grande. Grande, however, was even less expert as a gymnast than Sultan : even undisturbed by Rana, she did not achieve more than a short and feeble swing. As she turned away, Rana solved the problem with an impressive sweeping parabola. She was much better in this respect than Grande, who, like Tschego, left Mother Earth seldom and unwillingly.

(5)

An experiment, practically a reversal of the "tool-using," consists in placing a movable object across the path of the objective so that the problem can be solved only by its *removal* : it is impossible to "go round it." In comparison with the cases already described, in which implements have been used, this removal of an obstacle appears to the adult human being extremely simple ; one is inclined to say, before the test : "here is something the chimpanzees can do at once." To my astonishment this estimate is not correct.

The obstacle used in all cases was a box : the somewhat heavy transport cage of Konsul, which was a familiar object to all the apes, and had been used as a footstool by Rana, Sultan and Grande,

(March 19th.) The box was placed in the barred room in immediate contact with the bars and standing on the smaller end so that it could easily be knocked over. Outside the bars, and immediately opposite the centre of the box, the bananas lay on the ground ; they could be reached at once with a stick, if the box were pushed aside or even knocked over.

Sultan's first actions were not clear : he seated himself on the box and tried vainly to reach the objective with the stick. Sometimes he shook the box a little. Finally he lost hold of the stick, which fell outside the bars, and no other was available. Then Sultan actually took hold of the box at one side and pushed it a little away from the bars, so that he could easily have reached the prize. But he walked off without paying any attention to it. The test was broken off here, as Sultan appeared, from its inception, indifferent and indisposed to take trouble. A little later, after the box had been replaced at the bars, the young animals were all let into the room. Only Rana shook the cage a little, but did not move it away, and presently Sultan, obviously excited by the competition, "took a hand," removed the obstacle, and pulled in the objective with the stick. He probably had no special interest in the matter on the first occasion, that is, he was not at all hungry. But it is unlikely that the same " excuse " held good for all the other apes.

(February 20th.) The scene was composed in the same manner, except that the objective lay outside the bars just in front of them, so that the use of the stick was not necessary. The smaller animals were led into the room, with the exception of Sultan. They all tried to secure the objective, by mounting the box and reaching from either side of it, and showed plainly the intensity of their interest. As they could not reach it in this manner, they began to climb and sit about with lazy indifference on the box. As even Rana made no effort to push the box away, we may perhaps interpret her shaking of it on the previous occasion, not as an attempted solution, but as the preliminary " first touching " or—and especially in Rana's case—first " smelling out," which is customary in their behaviour towards new and unfamiliar things. If the creatures did not learn more from watching Sultan cope with the problem, that is quite in harmony with the observations we have repeatedly made, to the effect that the chimpanzee has great difficulty in taking over solutions from others at second hand (we shall return to this subject in the second part of this work). Finally, just as we were about to break off the test, thinking our waiting futile, Chica suddenly struck the right solution. She propped her back against the bars beside the box, thrust sideways at it with all four limbs and, thrusting it back, grasped the fruit. She had to exert her full strength as Tercera had seated herself on the top of the box, and remained enthroned with impassive countenance throughout the proceedings. It is an open question whether she was too stupid to realize Chica's intentions or was the most cunning of them all. Perhaps she was a little of both, in a curious and remarkable mixture, but certainly also what, in humans, is politely termed " mentally indolent." In such situations we always found her " sitting on the box."

If Chica's achievement was performed in imitation of what she had seen Sultan do, then she certainly imitated the *substance* of his actions, not their *form*, for her movements in displacing the box were quite different from his, though both came under the category of " removal of the obstacle."

(February 22nd.) Grande, Tercera, Rana, and Konsul were subjected to the same test. Not the faintest trace of a solution was observed. In turn, they all stretched out vainly for the fruit, or sat dejectedly on the cage. Incredible as it seems when we consider their achievements in other directions, none of them grasped this simple solution in spite of the examples they had had. Finally Chica was permitted to join them ; she saw the objective, seized the cage forthwith, and flung it over (and completely over) into the middle of the room ; as she flung it Rana also took hold of it for an instant, and Chica secured the prize.[1]

---

[1] Her manner of removing the obstacle was quite different from the first occasion. I wish to stress this point for the enlightenment of students who have not observed chimpanzees carefully. What Chica did in this experiment was to *clear away the cage from before the fruit, not* to make this or that series of movements.

In the next test both Sultan and Chica were not permitted to be present. The cage had accidentally been placed in an insecure position and wobbled easily. Rana tipped it over towards the inside (i.e. the room) just as Chica had done previously, partly with her help (see above), and reached the objective. It is very probable that this solution was " borrowed "; it was not purely by chance that Rana touched the cage in the former experiment just as Chica was knocking it over. The insecure position of the cage, which shook at a touch, this time, perhaps, made it easier for her.

[(March 16th.) We observed that Tercera, confronted with the same difficulty, shoved a box out of her way.]

(February 23rd.) The same situation was set up for Tschego, except that we had to take into consideration the much greater reach of her arms. It was the first time that Tschego had carried out any experiment. For a long while her response consisted in useless stretching and groping towards the objective while seated on the cage. Finally, we put down a second objective outside the bars and nearer to them than the first, a trifle to one side, within Tschego's reach, but still strongly obstructed by the box. Tschego took hold of objective number two, but did not respond to this assistance. She crouched beside the box, facing the bars. For some time nothing happened. Then, however, some of the smaller apes approached from outside the cage —they were permitted to do so as experimental stimulation —and endeavoured to approach the prize. Each time, though, Tschego repulsed them with threatening gestures, wagging of the head, stamping with her feet, and pawing the air with her great hands : for she regarded the objective as her property though it was beyond her reach ; otherwise she would not have menaced the little creatures, with whom she was generally on the best of terms. The youngsters finally gathered closely round the fruit, but the danger inspired Tschego ; she gripped the box, which was like a toy in her arms, jerked it backwards, stepped up to the bars, and took the fruit. In this case to know the *time* is of importance ; Tschego began to make efforts towards the fruit at 11 A.M., and succeeded at 1 P.M. If the little ones had not intervened, the test would have taken much longer.[1] And this was the case with the solution of a problem which to us appears so elementary, hardly more complex than the pulling of the objective by a rope already attached to it, which the ape does without hesitation. Another thing is to be noted in this procedure. The " obstacle " test was *not* solved either in the case of Tschego or the young apes by a series of imperceptible pushes involuntarily given to the cage in the act of stretching towards the prize. Quite the contrary : *during the lapse of two hours, Tschego did not move the cage one millimetre from its original position*, and when the solution arrived, the cage was not *shouldered* to one side, but *suddenly gripped with both hands*, and thrust back. It was a *genuine* solution.

The next day we repeated the test. The cage was placed exactly as before. Tschego perceived the objective, seated herself beside the cage in her former position, made one ineffectual effort to grasp the fruit, then seized the cage tipped it over backwards into the room, and took the fruit. Time : instead of the former two hours—barely one minute ; and the incitement of competition with the youngsters was no longer necessary. The *movement* by which she disposed of the cage *was quite different from that of the day before :* she

---

[1] This will show that results of value can be obtained only by the exercise of much patience and lapse of time. I have experienced more than once that success was attained in a *happy moment* after *hours* spent in vain.

did not repeat the innervations of yesterday, but *removed the obstructing cage.*

[A full month before this test, Rana showed remarkable behaviour in an experiment which on a cursory examination appears very like that just described. (January 25th.) A large cage, a good deal heavier than the one previously mentioned, was standing on free ground ; it was enclosed on one side by bars (and otherwise by wooden walls), so that it was possible to observe that it was resting with the door side on the ground. These cages—several of them—were standing about, and the apes often walked into them if the doorway was clear and not down on the ground. The objective was placed inside the cage, not accessible through the grating, so that the apes stretched their arms in vain between the bars. After some incredibly clumsy attempts to scrape the fruit towards her with a stick, Rana made quite unmistakable efforts to tip over the cage. If she had succeeded, it would have been easy to enter the cage through the doorway and secure the prize. But the weight was too much for her. About five metres away stood a similar sort of cage, with the doorway facing in the direction of Rana's laborious efforts. She suddenly stood still, approached the accessible cage, slowly entered it, turned round, and reappeared with an extraordinary expression of mingled stupidity and reflection ; then returned to the first cage and tried once more to overturn it, but in vain. I think it would have been impossible to observe this occurrence without the conviction that her sudden strange excursion into the other cage was the direct result of her efforts to turn the experimental object so that she could enter it. Later, in this account, I shall describe a procedure on Rana's part which somewhat resembles that just described and admits of only one interpretation.

Quite apart from this interlude the fact remains that

Rana had previously tried to overturn a cage in order to reach the doorway. This appears to be the same achievement as above, only, if anything, somewhat more difficult. We must try to see the two experiments in such light that the apparent contradiction vanishes.]

The results of these experiments were later confirmed on all occasions when the crux of a situation was the removal of an obstacle. The chimpanzee has special difficulty in solving such problems ; he often draws into a situation the strangest and most distant tools, and adopts the most peculiar methods, rather than remove a simple obstacle which could be displaced with perfect ease.[1]

We must, however, be on our guard against constructing our standard of values for these tests on the basis of human achievements and capacities ; we must not simply cancel what appears to us intricate, and leave what appears to us elementary in order to arrive at an ape's capacities (for, to an adult human, for example, the removal of an obstacle appears easier than the use of box or stick as tool, whereas to an ape, both present equal difficulties). We must avoid such judgments because the primitive achievements we are here investigating have become mechanical processes to humans. Thus the comparative difficulty of achievements may have been quite altered, nay, reversed, by the increased *mechanization* of these processes, the degree in which this has taken place being independent of the original difficulty. At the present time it is impossible to decide whether the processes which have become mechanical, and appear to us the easiest, have *originally* evolved most easily and, therefore, earliest. We can only judge what is originally easy, and originally difficult, by means of experimental tests with anthropoids and perhaps other apes, with children and

[1] The task is only facilitated if and when the object in question is moved by accident.

primitive peoples (for more advanced problems), and perhaps
also with imbeciles and mental defectives.[1]

[1] The example quoted on p. 39 shows that the results *sometimes*
entirely confirm the expectations of adult man. But we must keep
in mind that experience in the experimental test alone should be
allowed to decide.

III

THE USE OF IMPLEMENTS—(*cont.*)

HANDLING OF OBJECTS[1]

EXPERIMENTAL tests are not necessary in order to induce the chimpanzee to handle the objects of his immediate surroundings in a variety of ways. His large, powerful and flexible hands are natural links between himself and the world of things, and he attains the necessary amount of muscular force and co-ordination at an earlier age than the human child. His feet, although far from being " a second pair of hands ", can still be used in emergencies in which the feet of white races would be totally excluded. His jaws and teeth can also be very serviceable, as is indeed the case among African tribes and perhaps other primitive peoples, though to a lesser extent than with the anthropoid apes.

Though the chimpanzees under our observation developed very considerable procedures with these means, one can hardly say that these accomplishments were largely due to their captive state. The various objects ready to the captive's hand are hardly more numerous and diverse than the products of nature in the forests of the Cameroons. Moreover, a rag of cloth and a large leaf, a fragment of looking-glass and a small pool, are functionally similar, so that their use is almost interchangeable, and the presence of a few human " artifacts " makes little difference in this respect.

[1] Cf. Appendix.

I would rather suggest that confinement in narrow spaces and the resultant ennui, and the lack of need for long and arduous journeys, with their attendant exhaustion, are factors favourable to the use of the objects. But the response of the chimpanzee to these favourable circumstances will always be limited and determined by his own very pronounced natural proclivities. For I must most emphatically state, after a full acquaintance with chimpanzees, that it may perhaps be possible for the length of time taken by a circus " turn " and by beating or such means, to compel them to an action, to a habit, an omission, or a method of procedure which is not spontaneous and the natural anthropoidal response to the particular conditions ; but so to *weld* an alien nature into his own that the chimpanzee will continue to exhibit it when not under pressure, appears to me difficult in the extreme, and probably impossible. I should have the highest admiration for a pedagogic talent which could achieve such a result. It is a continuous source of wonder, and often enough of vexation, to observe how every attempt to re-mould his biological heritage " runs off " an otherwise clever and ductile animal of this species " like water from a duck's back ". If one is able to produce a—very temporary—type of behaviour which is not congenial to the chimpanzee's own nature, it will soon be necessary to use compulsion if he is to keep to it. And the slightest relaxation of that compulsion will be followed by a " reversion to type " ; moreover, while such pressure lasts, his behaviour becomes ugly by being constrained and indifferent to the essence of what has been demanded of him. It cannot be too much emphasized that no conclusions can be drawn from antics performed by chimpanzees on the stage, and under pressure of extreme force.[1]

Therefore, in describing not only the use of tools, but the handling of other objects, as may be observed daily, we need

[1] I remember having read such conclusions more than once, in the past.

have no fears : though the animals may be granted oppor-
tunities which they would not enjoy in their native African
forests, we shall always in these experiments be observing
the natural chimpanzee, and not any artificial product—
that is, so long as no compulsion is used ; and it is, of course,
to be assumed that no human pressure was employed in the
observations we are about to describe. When the animals
were quite unaware of being under observation, their behaviour
was the same.[1]

[This descriptive account does not refer to a limited period
but deals with the recorded events of fully two years, because
the chimpanzees are very subject to fashions, so that no
adequate idea could be conveyed except by an account covering a long period of time.]

The everyday handling and treatment of objects on the
part of the chimpanzee comes almost entirely under the
rubric " *play*." If under the pressure of " necessity," in the
special circumstances of an experimental test, some special
method, say, of the use of tools, has been evolved—one can
confidently expect to find this new knowledge shortly utilized
in " play," where it can bring not the slightest immediate
gain, but only an increased " *joie de vivre*." And on the
other hand one or other of the manipulations undertaken
in the course of play can become of great practical utility.
We will begin with a form of play that possesses this quality
of utility (somewhat overrated in Europe) in a high degree.

Jumping with the aid of a pole or stick was invented by
Sultan, and first imitated, probably, by Rana. The animals
place a stick, a long pole or a board upright or at a slight
angle on the ground, clamber up it as quickly as possible
with feet and hands, and then either fall with it in some
direction, or swing themselves off from it in the very instant
that it falls. Sometimes they spring to earth again, at other

[1] Sometimes even much more interesting.

times on to a grating, beams, the branches of a tree, etc., often to a very considerable height. And at first it was certainly not circumstances that "forced the leap upon them." They could have "got there" much more easily by walking or climbing. Also the landing-stages they selected seemed to offer no special attractions, so that when we take into account the constant repetition of this performance, we can only conclude that it is done out of the wish to *jump and leap per se*, just as children walk on stilts "for fun."

But very soon this form of play developed into the regular use of a tool. (Jan. 23,'14.) Sultan made the attempt to reach the objective (in the course of an experiment) in vain, as it was hung too high for him. He leapt straight into the air from the ground several times, and in vain; then he seized a pole that lay in his vicinity, lifted it as though to knock the prize down, and then desisting, put one end of the pole on the ground beneath the objective, and repeated the "climbing jump," as above described, several times in succession. His movements had a certain playful and sketchy character, as though to say: "It won't be any use!" and so it was not. On the next occasion (February 3rd) he was more resolute and more fortunate; he approached a solid piece of plank, so heavy that he could only just cope with it, placed it in position and started climbing and jumping off. Three observers who were present maintained that he could not possibly reach the prize in this manner, and on three occasions the treacherous plank fell over before he could reach the top, but on the fourth trial he succeeded and tore down the fruit.

The use of the jumping-pole spread to Grande, Tercera and Chica and even to the heavy and clumsy Tschego, but skill and success with it varied greatly according to their individual ability. After some time Chica was easily first: she "jumped off" with the aid of short sticks and boards, and presently with a pole of over two metres long, which had appeared from somewhere. By its aid she could reach anything that was not more than three metres above the ground. (Cf. Plate II, and notice that the stick is placed in the ground without other support, and that the house is several yards behind).

Later on, wishing to see how far her capabilities extended, I presented her with a bamboo over four metres long. She immediately showed complete mastery of this tool or toy, and climbed at frantic speed to a height of over four metres before the pole fell over. She herself at that time was not quite one metre tall, when drawn up to her full height. For certain reasons she had to be separated from her beloved toy for some time during the daytime ; but in the evening, when she entered the playground where the bamboo lay, she repeatedly interrupted the (to her immensely important) business of a *meal*, in order to seize the coveted treasure and " just once " snatch a hasty jump.

[Of course this clever trick was only possible as a result of experience in placing the stick and controlling her own muscular efforts, in order not to lose balance before she had completed her climb. We must compare this to the achievements of a human gymnast : Chica has a " feeling for it ". The draw-back is obviously the violent impact of a headlong fall from five metres on to a hard piece of ground. Chica often inspects and touches those portions of her body which have borne the brunt of the fall, and walks away with slow and subdued gait ; but, thanks to her incomparable skill as a tumbler, she received no serious injuries. There was no " training " whatsoever about this, either : *my* part in the matter was solely the gift of the long bamboo. The jumping off was invented, introduced, further developed, and utilized to solve problems in the tests, by the chimpanzees themselves. I could not have altered it if some day they should have tired of it—only tried to find a longer piece of bamboo for Chica.

Imitation of human beings is excluded in this case. For, although acrobats may perform the same trick, there were none such in Tenerife and the ordinary pole-jumping of expert human acrobats is something quite different—and not customary in the surroundings of the animals.]

A later modification of this accomplishment arose after the apes were enclosed in a narrower room with a low roof of extra strong wire-netting. The pole jump was executed to the height of the roof, the roof was seized but the pole was *not* relinquished, but used as a sort of gigantic office-stool. The obvious sequence is to clamber along the roof by means of the hands, and manage the stick simultaneously with the feet and still in a sitting position ; but this " game " was not observed very often.

If the pole or board were thicker at one end and the weight unevenly divided, a man would always place the larger and heavier end on the ground. But even in the case of Chica it is not evident whether or not she attaches any importance to this. As a rule the heavier extremity rests on the ground, but the other position can also be observed, and it is quite possible that the more advantageous position is only used more often because it entails less effort. When there is but little difference between the two ends, the chimpanzee certainly takes no heed of it. When we see him, or her, using the pole or plank with what we consider the wrong end uppermost, and yet springing with ease, we are inclined to think the error immaterial. But further experience teaches that it is *fundamental*.

[Rana makes an unprepossessing impression when she prepares to " take off " for a high jump, and the stick is too short. The other apes would look up and then throw away the pole, or at most make one attempt and then give it up.

PLATE II. CHICA ON THE JUMPING-STICK, RANA WATCHING

Not so Rana ; she props up the stick, attempts to climb it, stops, turns the stick as though that would make it grow longer, lifts one leg, lowers it again, and repeats this process a number of times, a picture of confusion and helplessness. Finally, as a rule, she squats down, lets the stick fall to earth, and stares vacantly around.

Dogs as a species are on a different enough level from chimpanzees, but as the chimpanzee has the wide range of individual variation corresponding to his high stage in evolution, nature has given to some individuals of this species an occasional expression of incredible and absurd stupidity. No dog could ever look so foolish ; his features always have a certain " neutrality ", and so no dog could show either the alert intelligence of aspect peculiar to very gifted individual anthropoids. Rana's stupidity is conspicuous as she is not only " *bornée* " but also extremely assiduous, and thus continually " shows herself up ", while Tercera, who only rarely takes part in an experiment, has succeeded for years in remaining something of an enigma. It is a significant fact that Rana could never find a real constant playfellow, except little Konsul, whom she more or less mothered as long as he lived. Her fellows had " no use for her ", and Tschego treated the hopeless Rana like a stupid clown.]

The stick is a sort of general tool in the chimpanzee's hands ; it can be turned to account in almost any circumstances. When its use has become common knowledge and property, its functions extend and vary from month to month.

Any object that lay beyond the bars and attracted attention but could not be reached by hand alone, was pulled in by means of sticks, wire, or straws. When the rains were over and the favourite food (green stuff) no longer visible on the playground, herbs and grass still grew outside before the wires. Then a stick was forced through the meshes, and the adjacent plant or bush pressed against the bars, so that the

F

foliage was within reach ; hours were spent in this occupation. But the wire-netting was old, and the pressure on the diagonal stick tore open a rent through which hands hard as iron and leather could get a grip and then tear holes, large enough to admit the chimpanzee's whole body. For a long time the apes had given no sign of dissatisfaction with their confinement, but after this discovery, we found they appreciated a little " voyage of adventure " keenly. I do not consider that their efforts were directed to that end from the first, but if there was a tear in the netting, however small, we could easily see from afar, by the suspicious assembly of animals at that point, that it was no more a question of green stuff merely. Even if we had not so frequently caught them in the act, the iron or wooden stave thrust through the torn netting would have borne witness after their flight.[1]

The use of sticks evolved in a similar way in the following case. The tank, which received the waste water used to wash out the cages, was closed by a thick wooden lid and iron bolts. But there were cracks, and it became a perfect mania with the apes to squat beside the tank, armed with straws and sticks which they dipped in the foul liquid and then licked. Of course things would have been much easier if the lid had been removed, and, either because it moved easily under the groping hands, or because it was easy to understand the situation, *this* obstacle was removed " early and often " either with the naked hand of a creature whose strength is capable of bursting open an iron bolt bedded in cement, or, later, as we increased the solidity of the structures, prized open by means of the stick, which had formerly functioned as a spoon, but was extremely popular on its promotion to the diginty of *lever*. The chimpanzee uses a

---

[1] These escapades were attended with no danger from the apes. They were so naïve that, if we scolded them on meeting them outside, they forthwith returned to their abode.

lever in exactly the same manner as a man. Of course the apes have no knowledge of the abstract relations between force, work, etc.,—the factors governing the aspect of leverage in physics—but the carrier, who lifts his car after a wheel is broken by placing a lever under it, hardly knows more about physics. There must be a kind of purely concrete and practical " sense " of such elementary implements, arising from the optical and motor functions of naïve people or the apes that can be relied on (within certain limits) quickly to seize and retain their application (cf. for instance, the case of the jumping-pole). Later on, when the apes shook the lid of the tank in vain, they did not trouble to dip the stick through a crack like a spoon, but at once tried to prize the lid open. Only when the lid would not yield to their efforts, did they being to dip again.

[The forcible opening of this tank became one of the greatest " fashions " observed among the chimpanzees. It took a long time before they were bored by the sport. It would be erroneous to assume that the main attraction to the chimpanzees was the dirt in the tank. The possibility of thoroughly and methodically " reducing any object to its component parts " must have been quite equally attractive. If a chimpanzee is placed in contact with anything breakable, the result is always fragments—and not because of the animal's clumsiness at all : *for he has no peace* until the remains and debris either are not worth further demolition, or do not permit of it. Possibly, however, it is only his superior muscular strength that enables the ape to outdo human children in this respect.]

Straws and twigs are also used as spoons in pure " play " during meal-times, when the animals have free access to drinking water. When their first thirst has been quenched in great gulps, one of the animals will take a straw, dip it into the water and suck it : this may happen twenty times in succession.   Once when a mouthful of red wine was poured into the water of the drinking-bowl that they all shared in common, they stooped unsuspectingly to drink, but after the first mouthful they paused for an instant.   Then one of them began to dip and spoon up the new mixture with a straw, and three others immediately followed his example with twigs and rags of cloth : it must have been too strong a brew for their usual hearty gulping.   There was no imitation of human beings in this case : for at that time they could only by chance have seen a human being using knife, fork, or spoon with his food.   These luxuries are not used by the natives in the neighbourhood of the station.

[The straw was used in quite a different manner by two of the animals (Grande and Konsul) on some occasions during the consumption of food.   When their hunger is not too acute, all the chimpanzees are in the habit of making a pulp or mess out of the fruits they eat (bananas, grapes, figs, etc.) ; this they roll to and fro in their mouths which are very flexible and often alarmingly expanded, or they contemplate it on their lower lip which is thrust far forward, or take it in their hand and look at it with satisfaction before replacing it in their mouth.   Both of those animals had a frequent habit or " fashion " of collecting straws and masticating them along with fruit pulp, but producing them again with great care, in a sort of knot or lump, when they had swallowed the edible part of their meals.]

A " halfway-house " between a spoon and a weapon of the chase is a twig or a straw used to capture ants.   At the height of the summer a small species of ant forms a perfect plague in Tenerife.   Wherever they pass, they form wide streams of moving brown, and this stream also pours itself along the beams around the wire-netting encircling the play-ground.   The chimpanzee has a special taste for acid fruit, which he prefers to all others ; and so he also relishes formic acid. If he passes close by a board or beam covered with ants, he simply rolls his tongue along it and gathers them in ! On the beam around the wire-netting he could not pursue this primitive method, as the ant-stream was *outside* the wire-netting. So, first one of our animals, then another, and then the whole company, began to stick twigs and straws out through the meshes and drew them in immediately, covered with ants, which were promptly devoured. The second time the saliva adhering to the twig or straw was immensely helpful, as in the fervent heat of summer ants seek any speck of moisture, and run in crowds over the damp straw ; indeed, they often had this advantage the first time, as chimpanzees generally lick the tip of a stick or blade of grass before using it for anything, just as some people do pencils. There can be no doubt whatever as to the meaning of the animal's procedure in this matter. They allow one to observe them at the closest quarters, their attention being entirely absorbed by the procession of ants. The straws are held for some seconds motionless amid the densest throng of insects, and when they are swiftly lifted to the mouth and pulled out again, not one ant remains on them. Nor is anything spat out, as is invariably the case when there is an unpleasant taste, e.g., when medicine is smuggled into the food. Probably the " play spirit " was as strong and stronger here than the special relish for ants ; for there were enough places available where they could be enjoyed with one flick of the tongue, and, when " fashion " had taken a different turn, the most profuse hordes of ants were simply ignored. But while the fashion lasted, all our animals were to be seen, squatting side by side along the ants' pathway, each armed with straw or twigs like anglers on a river's bank.

From time to time the use of sticks for digging becomes the fashion. Probably the only incentive necessary here is a stick with which the ground can be prodded. Digging gives more pleasure when the ground is damp than dry, and, when once begun, is carried on with enthusiasm and endurance till the place is full of big holes.　The chimpanzee holds the digging stick in a variety of ways : he by no means limits himself to the use of his hands, but, in hard soil, he thrusts downwards, using his great muscular development in nape and mouth, gripping the stick between his teeth.　Later on, the foot was used just as often ; the sole, which is extraordinarily tough and insensitive, is pressed hard against one end of a stick held diagonally in both hands, and drives it into the earth.　This was not an occasional or accidental method : Tschego nearly always dug thus.　The use of the foot as a hand—grasping the stick with the big toe—was much rarer. As will be realized, we are here very far on the way to the " digging-stick " known to our ethnologists.[1]　But this resemblance becomes more suggestive, when we take into consideration the fact that, before the " digging-stick " fashion first appeared, the apes had long been in the habit of digging and scraping up roots for food, when green things had withered in the summer heat.　They had used their hands first, showing great pertinacity, but when they began to use sticks, their progress was quicker and they could reach a greater depth ; so we cannot be surprised that the laying bare of roots became a favourite pastime.　Again, it was the oldest of the animals, Tschego, that distinguished herself particularly in this *rooting*, mainly owing to the enormous power of the limbs, jaws and teeth which wielded her stick.

　I do not wish to affirm that a chimpanzee picks up a stick and says to himself, as it were (for speech is definitely beyond his powers) : " All right, now I'm going to dig roots ! "　But no observer can be in any doubt that, in the course of desultory digging, the discovery of one root is followed by the definite

　　[1] The foot-pressure method is not copied from us. " Spades " are unknown in Tenerife.

search for more, by digging further, because he is already in the habit of scraping for roots by hand, and now finds that the stick is both more rapid and more convenient. Chimpanzees frequently search for something not already to hand. I made several investigations on the locality-memory of these animals, by burying fruit before their eyes. They not only found the exact spot later, but they looked in the vicinity, digging about before and after finding it (like a human being, for about half an hour, in the hope of finding more. The search for something differs from the mere digging game, very obviously, by the concentrated and eager attention, the hasty scraping and rummaging in the soil, the sifting of loose earth, the lively interest shown in each other's plots, etc.

Objects which are interesting, but unpleasant to handle, are forthwith approached by means of a stick. Nueva was once sitting beside me before a pile of twigs, which I set alight in order to observe her behaviour. She looked at the flames with a moderate degree of interest for a while, and then grabbed at them with her hand, which she at once hastily withdrew, only to seize a handy stick and poke it into the fire.

If a mouse, a lizard, or any small creeping creature wandered on to the playground, while the chimpanzees were there in occupation, it became a source of excited interest, but the big animals hesitated simply to capture it by a sudden snatch with the naked hand. It was grotesque to see the apes stretch their hands out with the intention of seizing their prey, with fingers pointed, and then, at once, draw back quickly. A firm grip of small rapidly-moving animals or reptiles seems as impossible to them as to many people among the human race. Every movement of the poor fugitive is followed by nervous gestures, half of fear and half of defence. We human beings, in a similar case, automatically project our elbow, probably because the unpleasant "tickling" sensation of touching the creatures would be less perceptible there than on the hand. Tschego does just the same. One start forward on the part of the lizard—who habitually moves in small runs—is enough to make the great ape jump and shrink, with a defensive forward thrust of her elbow, while her eyes close as though to ward off a blow. Of course, here too, a little stick is much better than the elbow, for the ape can poke about with it, and, in dealing with these little intruders, the chimpanzees always do use sticks, though in a very nervous manner. When the victim moves quickly in the direction of any ape, the fumbling stick it is true, becomes a weapon, and if the stranger does not succeed in escaping, it is killed after all, not in any spirit of deliberate cruelty, but in sheer excitement of the pursuit and capture. Between the tentative taps and pokes, which they simply cannot help giving any new and strange object, and their automatic defence when the creature moves, it receives so many blows and thrusts that it eventually dies.

Chimpanzees often smear themselves with excrement, either their own or their comrades'. Here a curious point of contrast must be recorded. I have only observed one of this species who did not take to coprophagy during captivity (Koko, to wit). Nevertheless, if one of them steps in excrement, the foot cannot as a rule tread properly after— as would also be the case with us. The creature limps off till it finds an opportunity of cleansing itself, and that by preference *not with its hands* (though it may, a few minutes before, have lifted similar filth to its mouth, and refused to leave go, even when it was given sharp blows on the hand), but with twigs, rags, or pieces of paper. As the ape's behaviour and expression are plainly indicative of discomfort, there is no doubt that filth *on the surface of its body* is disagreeable to it. This always happens when the body is soiled in any way.

The naked hand is not used to remove it, but it is rubbed with rags, etc., or against the wall, or the ground[1].

Such irrationalities and inconsistencies also appear very frequently when we study ethnology. In considering these cases we must be extremely wary of all strictly intellectualist interpretation of customs and institutions where there is any strong emotional content or background. The repulsive example quoted from the psychology of anthropoids is only a very vivid instance of the possibility and persistence of behaviour logically contradictory for the observer.

I had an excellent opportunity of observing the rapidity and ease with which chimpanzees have recourse to sticks, when they cannot well tackle any object, when, on the first occasion of their lives—at least as far as we can judge—they were brought into touch with high-power electricity. One pole of an induction coil was fastened into a wire basket full of fruit, suspended from the roof, the other was connected with a wire-netting on the ground. I have never seen, in such a short space of time, so many human reactions and expressive movements on the part of the chimpanzees. The starting back at the first shock, the cry of surprise, the cautious second attempt with constant jerks backwards, long before there is any possibility of the current passing through their body : the violent wagging of the hand in the air, especially after a strong shock, which exactly resembled the behaviour of a human being who has inadvertently touched a hot stove : these successive reactions to observe how many of our auto-

[1] To speak more generally, the outer surface of the body is treated with implements by preference. If water or oil are poured on an ape, it will rub off the moisture against a wall, the trunk of a tree, or (by preference) will wipe itself with straw, or a piece of paper. Blood from wounds is wiped off in the same manner ; one can often see them dabbing at it with leaves, which are generally moistened with saliva, or touching it with straws. After Tschego had attained maturity, it was patent that, on the occasion of every menstrual period, she dabbed at the blood with paper and rags, to wipe it off. When the skin or an inaccessible shoulder-blade itches, a stone or potsherd is used to rub and scratch the place.

matic responses must have been formed in the dark ages, in
which the primates originated.

As in the case of Tschego and the lizard (see above) the
chimpanzees of many thousands of year ago, must have
leapt back from the touch of any stinging animal or insect,
with the same gestures and reactions as we humans from a
high-power electric current. Possibly, further research will
establish this also in the case of smaller species of apes.[1]
But we should probably not find that these latter creatures
had recourse to a stick after their first unpleasant experience,
in order to extract the fruit without coming into direct contact
with the dangerous object. One after the other, all our
chimpanzees did this. And at first they succeeded, being
armed with staves of wood. But the basket eluded them,
as it swung to and fro on its cable. Then, in their eagerness,
the apes seized thick wires and iron bars, and, of course,
received one electric shock after another. They became
very angry, but only Tschego, who had not given up her
wooden stick, gave battle to the unknown. Standing at her
full height, she belaboured the basket so long and hard that
it whirled through the air, and finally fell down. An hour
later, the apes were still tentatively stretching their hands
towards the now innocuous wire-netting round the fruit,
and drawing back before they touched it, even after having
frequently taken some of the fruit with impunity.

In this case the stick was obviously used as a weapon,
for Tschego was visibly incensed and struck out blindly—
in strong contrast to the first careful efforts to extract the
fruit from the wire-netting. But this use of the stick is the
result of special circumstances. Otherwise it was only used
aggressively in the course of " play "—but that often when
it became the fashion.

[1] I do not hesitate here to " digress ", if, by doing so, I can give a
more lifelike portrait of the chimpanzees.

In the earliest days of my time at the station, the chimpanzees would often approach me in a threatening manner. But I subsequently realized that none of these "attacks" was probably meant seriously. Grande, whose incalculable temperament was always highly excited by any new-comer into her surroundings, frequently advanced towards me like a sabre-swinging ruffian, with bristling hair, blazing eyes, waving arms, and a stick which greatly enhanced the effect of this display; but I could only have supposed that she actually meant to attack me, if, and while, I was still ignorant of the chimpanzees' habits. The sight of any stranger would excite her to the point of such a demonstration, but it seemed to be only a "bogey-man" show. For it never occurs to her to make the game into real warfare. If she is quietly ignored, she waves her stick and stamps up and down for a bit, but finally she gives a dab at one with her empty hand, and gallops off; the battle game is over. It is the same between ape and ape. If one of them takes a stick and approaches another in a bellicose manner, or hits or thrusts at him with it, *that is certainly mere play*. If the other animal should also take a stick—as happens sometimes but not often—and threaten, or thrust with it, that is also definitely play. But if a misunderstanding arises and the game becomes serious, the sticks are flung to the winds, and the apes fall on one another with hands, feet, and teeth. It is quite easy to distinguish playful contests from real ones by the pace of the proceedings all through. The brandishing of sticks is clumsy and comparatively slow, but if a chimpanzee "means business," his rush is lightning-swift and leaves no time for stick-wagging.

[When anyone on the outer side of the bars is to be annoyed —and it is one of the chimpanzees' choicest pleasures to tease each other or third persons—it is done by creeping cautiously up to the wire, and suddenly springing against it.

But apparently much greater amusement is derived from thrusting a pointed stick at the legs, or into the body, of the unsuspecting victim. Grande again is mistress of this un-pleasing art ; spectators, dogs, and fowls are stabbed when-ever an opportunity occurs. Why? Only street Arabs who ring bells at strange front doors and then run away, or do other such things, can perhaps give the true answer.

During the weeks in which this record was completed, it became the reigning fashion to stab the fowls. The method of procedure is so characteristic that it deserves a full de-scription. I wish to state categorically that I have observed all the proceedings detailed below on several distinct occasions. When the chimpanzees are eating their ration of bread, all the fowls of the neighbouring estate assemble round the bars, presumably because crumbs sometimes fall through the wire meshes. As the chimpanzees take an interest in the fowls, they are in the habit of taking their meal close to the bars and gazing at the poultry, or frightening them away by a kick against the netting. Three forms of play have developed out of this, which I should not have believed possible if I had not observed them all repeatedly with my own eyes.

1. Between two mouthfuls of bread, a chimpanzee will sometimes hold his slice between the meshes of wire ; a hen approaches to peck at the bread, but before she can do so, it is pulled back again. At one meal this joke will be repeated about fifty times. It is quite unmistakable. If one of the apes has no fowl in his immediate neighbourhood, he bends sideways with the bread in his hand and thrusts the bait towards one through the netting, and waits. Perhaps even the fowls would learn wisdom after a number of deceptions if at least *one* of the apes did not carry the game even further.

2. Rana, the most stupid of all our chimpanzees, really feeds the hens intentionally. There is no doubt about this. Suddenly, in the midst of the " tantalus " game, in which

she also takes part, she continues to hold out her slice, so that a hen can, and does, take several pecks at it in succession ; she gazes with lazy benevolence at the fowl. As she must feel each peck on her hand, and directly contemplates the fowl, holding the bread in position for it—until she herself wants to take another bite—one can really only use the term " feeding the fowls." Some persons regard all the higher animals, and especially the anthropoids, with a certain suspicion and irritation ; they may be pleased to learn that Rana's behaviour was a form of " play," and not of deliberate altruism, that it did not occur very frequently, and now is being displaced by a further modification.[1]

3. This third game is as follows : the fowl is attracted to the bars with a slice of bread, but in the very moment when she is about to peck it, the free hand of the same chimpanzee (or of another beside him) thrusts a stick or—even worse—a strong pointed wire into her feathered body. When two chimpanzees take part in this (one as baiter and one as thruster) there has certainly been no previous agreement between them ; circumstances decree that the momentary activity of each happens to suit the other ; they realize it and continue their " collusion."]

The habit of thrusting and hitting with sticks frequently develops into *throwing* them. In moments of great joy, e.g. when very good food has been provided, one animal often seizes another (or a human being, if present), shakes

[1] *Note added after this report had been completed.* The procedure was observed again. It occurred as has been described above, only that first Sultan and then Tercera were seen to take slices of bread, throw them to the fowls, and then eagerly watch them feed. The act of throwing here was quite different from that described below (aggressive throwing) : instead of hurling with menacing gestures, a quiet strewing of the bread, while contemplating the fowls with fixed attention. I repeat that I was not prepared for anything of this kind. But there cannot be the least ambiguity about the facts or the nature of the " game." It was not played with any food except the rather indifferent bread. I have described elsewhere (see Appendix) how the chimpanzees sometimes shared their food with one another.

him in his excitement, pretends to bite him, etc. In such cases Chica loves to take a stick and fling it vehemently at Tschego's ample back. This frequently happens in play. For some time Chica was in the habit of creeping up behind one of her companions as they sat quietly at rest—it was generally Tschego—armed with her stick, hurling it from the closest quarters, and then taking to flight. In addition to sticks, she used rolls of wire-netting, tins which lay among the refuse, handfuls of sand, and with special zest, *stones* of the most varied size and weight. A few days after we had taken over the station, Tercera climbed up one of the roof poles, armed with a stone, and aimed so well at one of the strange intruders that she narrowly missed his head. At that time, however, the throwing was far from expert ; as with human children, it took some time to attain the right co-ordination of eye and hand, and Tercera especially generally missed her aim by a wide margin and often lost grip of her missile before she meant to do so. In the summer of 1915 stone-throwing became so much the ruling fashion that I have sometimes counted more than ten throws in the course of one quarter of an hour, most, it is true, by the same animal, the gymnastic expert, Chica, who learnt to aim excellently, and expressed her skill with equal delight against her fellow-apes and us. Others, again, did little or no stone-throwing : I have never seen Tschego do so, although she was a dangerous creature (owing to her size and strength) and could only be punished by us, if she bit or otherwise offended, by having stones thrown at her. But instead of flinging these missiles back at us, she seized them, and bit them viciously.[1]

The smaller apes were also of necessity driven back by throwing stones, when they could not otherwise be reached :

[1] When the chimpanzee expects a stone to be thrown at him, he holds his arms over his head and turns his back. This protective attitude of the arms also follows any unexpected or terrifying noise, as of rockets or gun shots.

as, for instance, when they broke out through a hole in the roof ; but we took care not to hit them, only to frighten them. This consideration had the result that Chica collected the stones and returned them, with emphasis and aim. For, unlike the usage of sticks in thrusting and hitting, the throwing of stones tends to appear among chimpanzees as a form of attack, under the influence of rage. And, like ourselves, the chimpanzee does not only throw stones at objects that he can actually hit, but equally, for instance, at the bars, when a scolding human or a growling dog stand at the other side of them. His greatest urge at the moment, that is, to expend his accumulated rage in the direction of the offender, is thus gratified.

As we were sometimes compelled to throw stones at the chimpanzees, it is quite possible that our action may have influenced them in the same direction. But it would be a mistake to assume that the creatures were influenced by our action alone. In order to have a correct and adequate idea of the chimpanzee's use of tools, we must take into consideration the following cases.

Tschego did not throw stones. But when she was scolded, we could sometimes observe her stamping indignantly to and fro, throwing her head backwards and forwards, and not only shaking and clawing with her long arms in the direction of the scolder, but also seizing handfuls of grass and herbs, and tearing at them till the bits were strewn round her. If she had her blanket with her, she dashed it furiously on the ground, but always these gesticulations both physically and psychically, were partially directed towards the enemy, as were also the manipulations of the grass and herbs. One could not yet exactly use the terms " throw at " or " strike at ", but the creature was obviously *approaching* the use of a weapon. The excitement which expresses itself by throwing and hitting the movable objects in the ape's vicinity has naturally a tendency towards and against the object of anger. But I think it highly improbable that these *forms of expression* of anger are influenced by the contact with human beings. In such a state of primitive emotional excitement, all not *inherent to the chimpanzee's nature* is certainly completely discarded. The stone-throwing by the younger apes, which we had occasion to experience at first, also looked more like an explosion of anger than a deliberate attack with a weapon. It is quite in keeping that this explosion should be in the direction of the stranger : he is the " object of emotion ".[1]

Violent anger is however not the best medium for the observation of the very general behaviour here discussed. Weaker emotions, of longer duration than passing fits of wrath, are better able to evolve all their latent possibilities.

A chimpanzee was locked up alone in the cage. His companions did not come immediately to embrace and comfort him through the apertures of window and grating, in response to his howls and whimpers. He stretched his arms imploringly towards them, and, as they did not yet respond, he stuffed straw, his blanket, anything he could find between the bars or waved it in the air but always in *the direction of his mates*. Finally, in the extreme of distress, he threw one of his available pieces of property after the other towards the objects of his grief and longing.

Sultan was isolated and had to fast a little in the interests of science. He sat moaning in his prison while the rest devoured their food. Presently he concentrated his laments and entreaties on Tschego, who squatted near him, armed with a huge bunch of bananas, and who, on other occasions, had risen and approached to share her superfluous goods with him. He howled and held out his arms towards her.

---

[1] Since then (in 1916) I have observed all shades of explosive actions with various implements up to complete *armed attack* on an enemy in a small, but very lively, young orang-outang female.

She turned her back on him and he began to jump up and down and scratch his head. Still she did not come. He knocked upon the wall and the ground outside his cage, stretching his body towards her as far as he could reach. Finally, he caught up turf and straws, and angled in the air towards her, and then pebbles which he threw, not to hit her, but towards her, so that they fell near her.

Fruit had been placed outside the bars, as in many other experiments. But Sultan had no stick of sufficient length. He grasped vainly at the bananas and only relinquished this vain attempt after some time. His hunger increased ; he seized twigs and pushed them towards his coveted prize ; finally, he threw twigs, pebbles, blades of grass, and all available movable objects at the fruit, uttering plaintive cries the while.

In all three cases the animal does not by any means necessarily, or even generally, end by a burst of helpless fury. The creatures are not angry, but full of yearning and desire.

Accordingly, a wish or urge directed somewhere *in space* and, long unfulfilled, does inspire actions *directed towards its object*, with little attention to practical utility. Certainly Tschego might have been moved by Sultan's behaviour, but as an unattainable fruit is treated in the same way as Tschego, a purely *utilitarian* interpretation will not suffice. It follows that, under the influence of strong, unsatisfied emotion, the animal must do something in the spatial direction in which the object of his emotion is situated. He must somehow get into touch with this objective, even if not practically, must *do* something, even if it is only to hurl the movables in his cage towards it. All emotions directed towards objects in space have the same quality (cf. above anger).[1] It is not

[1] In fear, as is well-known, the direction of action turns definitely right about 180°. Many animals will rush along terrified directly in front of a motor, as if they had to follow lines of force, when a slight swerve to right or left would save them.

the place here to show how human children show the same reactions, and adults too, when habitual inhibitions yield to extreme emotion.

Chimpanzees make nests from early infancy onwards. The full-grown female, Tschego, did the best and most remarkable work in this line. If, in the evening, she found straw heaped in a pile on her sleeping-board, she would sit on it, bend a handful slantwise from the edge towards the inside, and seat herself, or at least put her foot, on the twisted end ; she would go on in this, working all round until she had formed a nest something like a stork's. The blanket was often roughly woven into it, though rather used as a cover on cold days. The nests which the young animals make are much more untidy and loose ; and there is usually no turning down of the edge. If, on any occasion, they take a little more trouble, their movements during the preparation of the nest are exactly like Tschego's and these by no means depend on the material used.[1] Nests are often built during the day for fun, or at least are sketched out ; a great many different materials, such as straw, grass, branches, rags, ropes, even wires are collected and used, not when a nest is needed, but the shapes are suggested when the material is available. It may be noted, for instance, that loose green food, whether twigs growing near the animals, or brought to them already cut for them to eat, is diverted on its way to their mouth and laid down, as it were, as the beginning of a nest. It cannot be said that this looks very intelligent : one is even reminded sometimes of some stupid habits of the chimpanzees described later, or of " fixed ideas " in human beings. In any case the behaviour of the same animals is quite different when they are clearly solving a problem. If

[1] The little ones can only have seen Tschego building a nest at the beginning, and then on rare occasions, when they had no opportunity of imitating her. I am of the opinion that they do not need an example at all,

the material under consideration is anything like stalks or twigs and if there is little of it, then we are confronted with the strange phenomenon that, whatever the circumstances, the first thing is never to make even a scanty support for the body to squat on, but to create a ring round the animal; this is always done first, and if there is not enough material, then the ring is the only thing that is made. The chimpanzee then sits contentedly in his meagre circle, without touching it at all, and, if one did not know that this was a rudimentary nest, one might think that the animal was forming a geometrical pattern for its own sake. If a tree with foliage be set up in the animals' playground, after a few moments the nest-making begins by bending in the branches, and pressing them down with the weight of the body (compare the above), as necessarily as a chemical reaction. Koko, the tiny one, who had been away from Africa and the example of other chimpanzees for months, when he could yet hardly climb a tree, would still, when three metres up, bend down the branches and begin building a nest at once. Thus, in this case, we may speak of the manifestation of a special and elaborated " instinct ", whilst chimpanzees do not, as a rule, show many other signs of behaviour which could be called by the name of this utterly unexplained riddle.[1] In any case this is not the species of animal on which to begin such investigations.

They are fond of carrying quite widely different objects about on the body in one way or another. Almost daily the animals can be seen walking about with a rope, a bit of rag, a blade of grass or a twig on their shoulders. If Tschego was given a metal chain, she would put it round her neck immediately. Bushes and brambles are often carried about in considerable quantities spread over the

[1] Birth and the care of sucklings in chimpanzees has been but recently investigated. (See *Ber. d. Preuss. Ak. d. Wiss.*, 1921.)

whole back. In addition, string and pieces of rag are to be seen hanging in long strings over their shoulders to the ground from both sides of the neck. Tercera also has strings running round the back of her head and over her ears, so that they dangle down both sides of her face. If these things keep on falling down, they hold them in their teeth or squeezed under their chin, but, whichever way it may be, they must have them dangling. Sultan once got into the habit of carrying about empty preserve-tins, by taking the side that was open between his teeth. Chica, the sturdy, at one time took a fancy to carrying heavy stones about on her back; she began with four full pounds and soon reached a ponderous block of lava weighing nine pounds.

The meaning of these things can be seen clearly in the circumstances and behaviour of the animals. They play, not only with the things they have hanging round themselves, but, as a rule, with other animals' also, and their pleasure then is visibly increased by draping things round themselves. It is true that one often sees an ape walking about alone and yet draped, but, even under these circumstances he is mostly impishly self-important or audacious, as on the occasion when a decorated chimpanzee, with all signs of being in the best of tempers, will strut about among his companions or advance upon them menacingly. The adult female, Tschego, was often thus festooned when she trotted round in a circle with several of the smaller animals, quite at ease, tossing her head up and down, her mouth wide open, and, unlike the occasions when she was preparing to attack, all her muscles relaxed. That the whole company was playing could not be doubted by anyone seeing them, marching in a circle, one behind the other, the big animal stamping its foot violently at every step, or every other step,[1]

[1] That Tschego begins to *elaborate* the march *rhythmically* when playing in a circle is certain, as also the fact that in other cases she cares more for the form of the movements of the body, while rhythm occupies a secondary place.

and the others exaggeratedly accentuating the marching movements. At the time of this fashion, Sultan also used to carry his tin pot chiefly in his mouth, whenever he went, in a playfully threatening attitude, towards one of his companions or towards spectators on the other side of the bars.

[I observed an example, one evening, in which there was no sign of mirth and play, when Tschego, who had not been brought to her resting-place at the usual hour, had to remain outside by herself, while it was getting darker and colder. She began to build a nest as a matter of course, but ever and again would feel uncomfortable and lope about the ground uneasily; finally she carefully picked up everything she could find in the way of dry leaves, twigs, etc., and put it on her back. All the time she was in the worst of tempers.[1]]

No observer can escape the impression that, apart from Sultan's tin pot and Chica's athletic block of stone, which leave strong room for doubt, the objects hanging about the body serve the function of *adornment* in the widest sense. The trotting-about of the apes with objects hanging round them not only looks funny, it also seems to give them a naive pleasure. Naturally we can scarcely assume that apes have a visual image of what they look like when dressed up like this, and I have never observed their frequent use of reflecting surfaces as in any way connected with their adornment; but it is very likely that primitive adornment like this takes no account of external effect—I do not give the chimpanzees credit for that—but is based entirely on the extraordinary *heightened bodily consciousness of the animal*. It is a feeling of stateliness and pride, feelings, indeed, which occur also in human beings when they decorate themselves with sashes or long tassels knocking against their legs. We raise our

[1] Cf. also Appendix; Reichenow, *Naturwissensch.* IX, p. 73 seqq., 1921.

own opinion of ourselves by self-contemplation in front of mirrors, but the enjoyment of our splendour is not dependent on the looking-glass, on visual images of our looks, or on any other visual impression ; *when anything moves with our bodies, we feel richer and more stately.*[1]

[Sultan, with his tin box in his mouth, often utters dark sounds, when he comes up towards other animals or men, sounds which echo still more hollowly in the empty box. If any doubt remains here that the acoustic effect of the box was noticed, and then used on purpose, it certainly must have considerable significance if some of the animals, in a state of excitement, will drag boxes, tin drums, and so forth, behind them on the floor, and rattle them tremendously when running up against any person, or even a wall (cf., for instance, p. 47). Grande, in particular, does this. In all sorts of circumstances, but sometimes without any apparent reason at all, she will put herself into quite a terrific state of excitement ; she gets up, her long fine hair streaming in all directions, so that she looks like a black powder-puff, seizes the box or the tin in her hand, stamps from one leg to the other with glittering eyes, and bent forward a little, setting her arms, and possibly the instrument also, rhythmically a-swing, and, after sufficient preparation, suddenly rages at other animals, men or walls. If the animal is walked over, the man has stepped aside, or the wooden wall has received a thundering kick, then her bristling fur settles down and her fury is spent. In this case the noise most certainly has some meaning ; for the same animal, when pretending to be cruel or terrible (in reality Grande is the most kind-hearted of souls), stamps loudly on a box, or, on the other hand, one can enrage her tremendously by making any noise, especially

---

[1] H. Lotze, *Mikrokosmos*, treats of similar points, only he talks of a top-hat, which the chimpanzees would certainly also make use of with joy.

by drumming on a box. As a matter of fact, the others will often get similarly enraged, but their fury never takes so dramatic a course as Grande's ; she, at her best, would rush along, hair bristling, arms rigidly stretched out like a ship in full sail ; she could succeed, in this guise, in giving a real shock to inexperienced observers (compare p. 86).]

A thing very probably related to this ornamentation is the carrying of all sorts of things between the lower abdomen and the upper thigh. The apes not only keep food there, when they need their hands for climbing or have too much to carry, but also, for no particular reason, pieces of wood, stones, rags, and all kinds of objects, which may give pleasure to the animals. Tschego, in particular, ran about for whole days with an object wedged in this spot which she never let go, even when she sat down to rest. Once, it was a red rag which she would not remove from her lap, another time, a round stone polished to smoothness by the sea. Once I gave her a photograph to see what she would do ; she looked at it for a while, felt it all over with her big fingers, and then put it in her " trouser-pocket ".[1] Once a thing is put there, it is hard to get it again. For instance, the animal looked after the smooth stone most carefully ; if she had squatted on the floor, she would press it firmly and carefully when she got up to change her place ; on sitting down again, she would feel it and turn it round ; on no pretext could you get the stone away, and in the evening she took it to her room and her nest.

To say that a game is again in hand does not altogether cover this case. For it is noteworthy that the region of the lap in a chimpanzee often signifies *much more than the geometrical* centre of the body, although the sex organs (in the female at least, and therefore in Tschego) are placed towards the back, so that they form more the termination of the back

[1] Cf. Appendix, p. 325.

than part of the lap. When, for instance, a small animal greets Tschego, it generally (there are a few other forms of salutation) puts its hand in the bigger animal's lap; if the movement of the arm does not go so far, Tschego, when in a good mood, and particularly when it is her friend Grande, will take the hand of the other animal, press it to her lap, or there pat it amicably. She will do exactly the same to us when she is feeling amiable, that is, she will press our hand against just that spot between her upper thigh and lower abdomen where she keeps wedged her precious objects. She herself, as a greeting, will put her huge hand to the other animal's lap or between their legs and she is inclined to extend this greeting even to men.

To see anything nasty in this habit, is to mistake the entirely innocent character of these animals. The animals in the zoological gardens, I am told, sometimes behave in a very ugly manner; the chimpanzees of the station are, without doubt, very dirty creatures in the usual sense, in spite of the care that they devote to each other's bodies, and are certainly very greatly " coprophagous ", but their sex-cleanliness could scarcely be greater; I have only seen little Koko, while hopping about in a fury, but on no other occasion, happen to masturbate in a manner which must have originated accidentally under these circumstances.

To this remarkable part of his body, then, its inmost spot, so to speak, when the chimpanzee squats in his usual negro fashion, does he clutch his property, and it is quaint to see how even the oldest animal, and the most difficult to influence, will always keep there any valuable he possesses, particularly rags and the like.

Once, when lumps of white clay were brought to the playground, the animals gradually began to paint, without any stimulation, and whenever afterwards they again got clay, the same game would begin. At first the chimpanzees licked the unknown stuff ; very likely they wanted to see how it tasted. The result being unsatisfactory, as usual in similar cases they wiped their protruding lips on the nearest object they could find, and, of course, made it white. But, after a while, the painting of beams, iron bars, and walls, grew to be quite a game on its own ; the animals would seize the clay with their lips, sometimes crush it into paste in their mouths, moistening it, and would then apply the mixture, make fresh paint, and daub again, and so on. *The point is the painting, not the chewing of the clay* ; for the painter himself, and the rest of them, when not too much occupied with their own affairs, are most interested in the result. Soon, as is to be expected, the chimpanzees stop using their lips as paint-brushes and, taking the lumps of clay in their hands, whitewash their objects much more quickly and firmly. Of course, they have not yet achieved more than big white blobs, or, when particularly energetic, getting a whole beam-surface whitened. Later on the animals should have been given other colours. Once Tschego very patiently painted her legs all over, but on her dark fur the effect was not satisfactory.

Through this moistening and painting, the animal's whole mouth, of course, soon becomes white. But while, when decked with all sorts of adornments, the animals become playfully self-important and pleased with themselves, when their faces are whitened, they behave exactly as usual ; so that the white face is a mere by-product of the animal's activity which it very likely does not realize itself.

The chimpanzee's methods of dealing with external objects has been sufficiently described for the present ; a few observations relative to plaiting, the use of keys, their behaviour with reflecting surfaces, are given in another connexion. Ethnologists will seize immediately upon what is of interest for them without further emphasis.

IV

THE MAKING OF IMPLEMENTS

In all intelligence tests of the kind applied here, one circumstance is always repeated : if one *single part* of the " solutions " in the proceedings we have discussed (e.g. the beginning) be considered by itself and without any relation to the remaining parts, it represents behaviour which, in the face of the task, i.e. the attaining of the objective, seems to be either quite irrelevant or else to lead in the opposite direction. It is only when we consider the *whole course* of the solution (or, as later, at least considerable sections instead of those parts) that this whole seems to have some significance, and each of the parts previously isolated takes on a meaning as *a part of this whole.*

[Only for *one* section is this not true, namely, the last, in which each time, as a result of all the preceding acts, the objective is merely grabbed. This section, of course, has significance even when considered alone.]

The above is not philosophy, it is not even a theory of the actual proceedings, but a simple statement with which anyone must agree who distinguishes between " significant in relation to a task " and " non-significant ", or " meaningless ", and who will consider examples objectively.

When a man or an animal takes a roundabout way (in the ordinary sense of the word) to his objective the beginning —considered by itself only and regardless of the further course of the experiment—contains at least one component which must seem *irrelevant ; in very indirect routes there are usually some parts of the way which, when considered alone, seem in contradiction to the purpose of the task,* because they lead away from the goal. If the subdivision in thought be dropped, the *whole* detour, and each part of it considered as a part of the whole becomes full of meaning in. that experiment.

If, with the aid of a stick, I fetch an objective, otherwise unattainable, the same applies. The act of picking up the stick lying near me, considered separately, and without reference to the other part of the performance, the use of the stick as an implement, is quite irrelevant in relation to the objective. It does not, keeping always of course this supposed isolation in mind, bring me any nearer to my objective, and is, therefore, meaningless in this situation. But, on the other hand, considered as part of the total proceeding, it has the significance of a necessary part of a meaningful whole.

The same reflections applied to other "indirect ways" (used figuratively) illustrate the same circumstances, and that is why we call them all "indirect ways".

Thus matters stand from the point of view of a purely objective consideration. How the chimpanzee arrives at his actual solutions in such cases is another question, which we cannot investigate until later. But the purpose of all further experiments is to set up situations in which the possible solution becomes more complicated. Thus the objective consideration of the course of the experiment in parts will show up, ever more clearly and in greater number, sections which, *taken separately, are meaningless in relation to the task,* but which become *significant again,* when they are considered *as a part of the whole.* How does the chimpanzee behave in such situations?

One group of such cases, which will be considered in what follows, we characterize by the phrase "making of implements ". For practical reasons this name is here used in a wider sense ; that is to say, every subsidiary action which treats an implement, which does not for the moment exactly fit the situation, in such a way that it *may* be utilized, is considered as the " making of implements ". This preliminary treatment, of whatever kind, forms the new constituent which, when isolated, has no relation at all to the goal, but which becomes significant as soon as it is considered with the rest of the proceeding, in particular the " utilization of the implement ".

(1)

It seems to be only a weak indication of such auxiliary action when Chica, chasing another animal in mock fight, sees a stone, wants to pick it up, and when it does not immediately come away from the ground, scratches and drags it until it becomes loose ; in the same instant she is on the chase again behind her adversary, and throws the stone at him.

A more important performance, which took place several times afterwards, was observed by Teuber (my predecessor at the station) : Sultan grabs at objects behind the bars and cannot reach them with his arm ; he thereupon walks about searchingly, finally turns to a shoe-scraper, made of iron bars in a wooden frame, and manipulates it until he has pulled out one of the iron bars ; with this he runs immediately to his real objective, at a distance of about ten metres, and draws it towards him.

[In this case it is pretty clear that the whole proceeding, part by part, contains *several* constituents which are meaningless when isolated. (1) Instead of keeping to his objective, Sultan goes away from it ; this is quite senseless when taken by itself. (2) He breaks up one of the station's iron shoe-scrapers, and this, taken by itself, has nothing whatever to do with his objective.

However, some additional remarks must be made about these two instances, in their actual occurrence: (1) The animal by no means strides away from the objective in the free, careless way which we are used to, in him and the others at times when they are seeking nothing, but goes away like someone who has a task before him. And here again, I wish to warn against anyone speaking of " anthropomorphism ", of " reading into " the animals, etc. . . . where there is not the least ground for such reproaches. I merely ask whether there is not a difference of behaviour between people strolling about idly, and people looking for the nearest chemist's shop, or for any lost object ? Of course their behaviour is different. Whether we can exactly analyze our total impression in both cases, has nothing whatever to do with the case. Now all I wish to state is that the two general impressions that are contrasted here *occur in chimpanzèes, exactly as in man* ; and it is these " impressions ", which are not at all " something that has been read into " the chimpanzees, but which belong to the elementary phenomenology of their behaviour, that are meant when we say, for instance, " Sultan trotted about gaily " or " he went over the ground looking for something ". If *this* is an anthropomorphism, so then is this sentence : " Chimpanzees have the same tooth formula as man ". So as to leave no doubt whatever as to the meaning of the expression : " Walking about searchingly ", I should like to add, that nothing is said therein as to the " consciousness " of the animal, but only as to his " behaviour ". (2) While occupied with the shoe-scraper, Sultan's activity is concentrated exclusively on *loosening one of its iron bars* ; but even when described more precisely thus, this action remains *irrelevant with reference to the real purpose as long as it is considered in isolation.*

At the time when Koko did not know any longer how to use his box (cf, p. 43), he came once to the same proceeding as Sultan, when the objective was hanging high up on the wall. Four metres away in front of the door was a shoe-scraper, exactly like the one mentioned. After a long look at the box, which did not lead to his making use of it, Koko turned away, espied the shoe-scraper, ran towards it, and began to pull at it with all his might, until the nails with which it was fastened to the ground, finally gave way. Satisfied, he then dragged the heavy board towards the objective. But he was startled on the way by a whistle near by, and dropped his burden, so that one could not say what would have happened further. But soon after, he again turned towards the board, stood on one of the edges running lengthways, and pulled and shook the iron staves as hard as he could, probably to tear them off; but as he was too weak, and also did not set about his task very efficiently, he had finally to stop his efforts.]

(17.2.1914) Beyond some bars, out of arm's reach, lies an objective; on this side, in the background of the experiment-room, is placed a sawed off castor-oil bush, whose branches can be easily broken off. It is impossible to squeeze the tree through the railings, on account of its awkward shape; besides, only one of the bigger apes could drag it as far as the bars. Sultan is let in, does not immediately see the objective, and, looking about him indifferently, sucks one of the branches of the tree. But, his attention having been drawn to the objective, he approaches the bars, glances outside, the next moment turns round, goes straight to the tree, seizes a thin slender branch, breaks it off with a sharp jerk, runs back to the bars, and attains the objective. From the turning round upon the tree up to the grasping of the fruit with the broken-off branch, is one single quick chain of action, without the least " hiatus," and without the slightest movement that does not, objectively considered, fit into the solution described.

At a repetition soon after, things did not go so smoothly, but that was not Sultan's fault. The branch was taken away in Sultan's absence, a new objective put down, and Sultan again admitted. He at once broke off another branch, but tried in vain to reach the objective with it ; for the branch, when the tree was sawn, had been bent in the middle. He pulled it back through the railings, bit through it at the bend and continued to work with the other half, but in vain, for now the implement was too short.

[In connexion with the biting through of the branch where it is bent, it should be added that all the small animals considered it a game to poke about with straws in holes and the joints of walls ; the feeble straws kept on getting bent during this performance, and the animals kept on biting them off to make them of use again, until eventually they became too short altogether. In the experiment, the biting off at the bend is both correct and incorrect ; *correct* because the half is really a better " stick " in the functional sense, *incorrect* because, even without being bitten off, one half would have served its purpose as an implement, had it been long enough.]

For adult man with his mechanized methods of solution, proof is sometimes needed, as here, that an action was a real achievement, not something self-evident ; that the breaking off a branch from a *whole tree*, for instance, is an achievement over and above the simple use of a stick, is shown at once by animals less gifted than Sultan, even when they understand the use of sticks beforehand.

Grande was tested the same day. She stretches out her arm to get hold of the objective, but all her efforts are in vain ; she cannot reach it. Finally she steps back from the bars, wanders slowly through the room, and squats close to the tree, chewing its branches for a while indolently. In spite of her "landing at the tree " thus and chewing at it, we do not get the impression that it has anything to do with the objective, which is no longer noticed. After waiting for a long time, during which no sign of a solution is given, the test is abandoned. I might mention the fact that Grande is older and much stronger than Sultan, so that she could have broken off a branch with the greatest ease.

Four months later (16.6) the experiment is repeated ; she has meanwhile become very much more accustomed to the use of sticks. The tree, consisting of three strong branches (without twigs) growing out of a thick trunk, is at the very back of the room, as far away as possible from the railings, and therefore from the objective (about five metres). Grande first of all tries to pull out from its metal rings an iron bar which is attached to a door of the room, as a temporary bolt. As her efforts do not meet with success, she looks about the room, lets her glance rest for a while on the tree, but looks away from it again, and then spies a strip of cloth quite close to the bars ; this she seizes and makes attempts to whisk the objective to her (compare above, p. 34). When the cloth is taken away from her, she rattles the iron bar again and, as it does not loosen, she looks round the whole room, and especially at the tree in the background ; she sees a stone on the floor, carries it to the bars and endeavours—in vain— to squeeze it through them ; obviously it is meant to take the place of a stick. After a further glance back, she at last marches toward the tree, leans with one hand on the wall, puts the other one, and one foot, against the branch furthest to the front, with one jerk breaks it off, refurns at once to the bars and attains her objective. In explanation we must add : the black iron bar, although *actually* much more firmly affixed to the door than are the branches to the tree, yet stands out *visually* better from the wooden door, *as a separate object*, especially as one end is bent in from the door to the

H

room. To " *see* " a branch of the tree, so to speak, *as a stick*, is much more difficult, and Grande did look at the tree twice, without this happening. From the moment she walks towards the tree, her procedure is just as determined and as " genuine " as Sultan's.

(1.3.1914) Tschego had used sticks as implements during the preceding days and even on the morning before the experiment to be described. A tree is placed about two metres away from the bars, and Tschego is then let into the room. She does not see the tree at first, but when her eye lights on the objective, goes as usual into her bedroom, fetches her blanket, stuffs it through the bars, throws it on to the objective, and tries thus to draw it towards her. For the blanket can be used in two ways, either of which might succeed : beating the goal towards her (compare above, p. 34), or pulling it towards her, after the cover has been thrown over it. The cover is taken away from her ; she seizes the tree and makes a great effort to squeeze it, just as it is, through the bars. When that does not succeed, she takes a bundle of straw in her hand, stretches out with it like a stick, and endeavours to pull the objective towards her. As the bundle proves to be too soft, and does not drag the objective with it when pulled along, she takes hold of the straw in the middle with her teeth, and at one end with her hand, and bends one half over the other, so that a bundle half as long, but incomparably firmer, a real sort of stick, is formed ; this she uses at once, and, since it remains long enough, again and again, with complete success. The whole proceeding, from the taking of the bundle of straw which is too soft, up to the use of the firmer one, is one cohering action ; it lasts but a few seconds. In this way a method of making implements has been invented that is different from the one expected ; Tschego did not, at any time, show any indication of breaking off a branch of the tree, but she clearly showed that she " had present " the *use of the stick* all through the experiment. The tree, by the way, was only a very small one, which Tschego could easily manage as a whole. This explains why she wanted to use this whole as a stick ; but the rough procedure by which she pushes it towards the bars, as if she could thus get it through, is, of course, not justified by the size of the little tree.

The next day the test is repeated ; the little tree lies in exactly the same place as on the day before at the beginning. Tschego uses a bundle of straw as a substitute for the stick, and, when it proves too soft, folds it double just as in the first test, making it stiffer. This time, even after folding it, it still remains too flexible, so she hastily repeats the proceeding, and the bundle, now composed of four folds, thus becomes extremely firm. But now it is too short, and Tschego tries to squeeze the whole tree through the bars. As, of course, she fails in this also, she returns to the straw, and, after many failures, finally sits down quietly. But her eyes wander and soon fix on the little tree, which she had left lying a little way behind her, and all of a sudden, she seizes it quickly and surely, breaks off a branch, and immediately pulls the objective to her with it. This proceeding has no relation to her former attempts to push the tree through the bars. While breaking off the branch, Tschego turns one side towards the bars ; the little tree does not touch them at all and is neither treated as a whole nor moved towards the bars ; nothing else is involved but just the *breaking off of the branch*.

[In this experiment, what is particularly worth noting is the fact that, for a long time, not the slightest sign was given of the expected solution ; when the branch is suddenly broken off, the proceeding goes on, without any " hiatus ", to the reaching out with the stick thus created : *both actions together make one united proceeding.*]

Koko attempted solutions of this sort from the very beginning, before we had undertaken such experiments with
him. On the first day of his stick experiment (compare
above, p. 33) he pushed the objective still further away by
some clumsy movement, so that he could not reach it at all
with his stick, still less with a stalk that was lying close by.
He therefore turned sharply to a geranium-bush that grew
to one side, seized one of its stems, plucked it, and advanced
with it toward the objective ; on the way he eagerly picked
off one leaf after the other, so that only the long, bare stem
was left ; with this he then tried—in vain—to pull the objective to him.   The pulling off of the leaves is both correct
and incorrect ; *incorrect*, because it does not make the stem
any longer, *correct*, because it makes its length show up better
and the stem thus becomes optically more like a " stick ".
We shall see later how important such optical conditions are
to the chimpanzee generally (during his attempts at solution),
for optical impressions sometimes gain the victory over
practical considerations.   There can be no doubt that Koko
did not pull off the leaves in play only ; his look and his
movements prove distinctly that throughout the performance
his attention is wholly concentrated on the objective ; he
is merely concerned now with preparing the implement.
Play looks quite different ; and I have never seen a chimpanzee play while (like Koko in this case) he was showing
himself distinctly intent upon his ultimate purpose.

Two days later, just as he has forgotten the use he made
of the boxes, Koko first of all reaches out his hand towards
the objective in vain ; then he looks round searchingly,
suddenly goes towards a bower thickly covered with foliage
(three metres from the objective), climbs up the bars of the
bower to a place where one woody branch stands out from
the others, bites through the branch (whose end is far in
among the bushes), first in one place, then again about ten

centimetres further on, climbs down again, runs underneath the objective, but remains sitting there without using the little stick he has brought with him, somewhat sulky, and sucking the wood. It is much too short. In this case the proceeding, from the time he started for the bower up to the return to the objective, is *one* united course. That a chimpanzee, after a glance at the distance to be traversed, will avoid the use of implements that are too small, can be seen again (compare above, p. 46). This, of course, provided that he does not act under strong " affect " (compare above, p. 88).

The animals themselves vary this proceeding and often produce strange solutions unexpectedly. Thus, embarrassed by the lack of a stick, their attention will be drawn to a piece of wire, partly separated and standing out from the coil, and thus seeming to have the shape of a stick. They will take great pains to pull it off altogether, and sometimes are successful. More frequently it happens that, in similar circumstances, they will go towards a box, a board, and so on, detach a piece of wood with their hands, feet, and teeth, and then use that as a stick. Cases where the animal works at the box or the board in play only and without taking the objective into consideration, until a piece of wood is detached which, *afterwards* (when the animal again turns to the objective), it may use as an implement—*such* cases must be excluded with severity, and I have made it a rule in this book that the least suspicion of this kind be considered enough to cancel the value of the experiment.

[One fact must be noted in reference to the breaking off of pieces from boxes, etc.: not everything that is obviously " *a part* " for man, is so for the chimpanzee. If a box be left with only its lid, and if this half consist of separate boards, the chimpanzee will not always behave in the same way, whichever way these " parts " are put together. If the separate boards are nailed to the box in such a way that they make an unbroken surface, i.e. the joints not noticeable, the chimpanzee will not easily see " possible sticks " there, even if he be in urgent need of them ; but if the last board towards the open half of the box is nailed in such a way that a small space or crack separates it from the next, it will be immediately torn off (cf. above). There must exist a kind of visual firmness, which makes the separating as an act of intelligence as difficult as the strongest nails would the actual pulling off of the board. For the optical firmness does not seem to affect the chimpanzee as if saying : " this board is firmly fixed "—but in such a way that he does not ever see any board as a " part." True, in principle, we do not differ from the chimpanzee in this, but, when in need, we will dissolve visual wholes of much greater firmness ; or, to be more exact, under the same objective conditions, visual wholes are probably more easily analysed by the adult human than by the chimpanzee. Man is more likely to see " parts ", when he needs them, than the ape.[1]

With some reservation, I should like to add to the above a still more far-reaching observation. It appears from our several observations that there need not be any actual connexion (in the technical sense) besides the (to the chimpanzee) apparent (optical) connexion, and that one object does not have to be " really " part of another, in order that the chimpanzee should regard it as if nailed to its surroundings, and not notice it at all as an independent object. If one places any implement—a window-grill, a compact table, etc.—in such a way that this object becomes as far as possible optically an integral part of its surroundings : for instance, if the table is placed carefully with one of its corners in the right

[1] If the object has been used frequently, the effectiveness of the purely optical is much diminished ; the same is true here as in the case of the influence of the distance of configuration factors relative to the positions of objective and implement.

angle of a room, and the flat window-grill against a wall, so
that it joins it completely : one will see the chimpanzee, in
search of implements, pass by this object, as if it did not
exist. The object is not hidden in this case, it merely forms
with its surroundings a perfect integral whole. I have not
been able to make many such experiments, for the simple
reason that I did not wish for the moment to hinder attempts
at solution, but, on the contrary, to favour them to a certain
extent by circumstances. Every experimenter will probably,
for the same reason, without much hesitation avoid putting the
box, for example (which may come in as an implement),
into the corner of a room so that it becomes visually part of
it. But if one is seeking a theoretical explanation, then once
we know what the chimpanzee can do, every experiment in
which we succeed in preventing the usual solution is of the
greatest importance.

The above obviously bears out M. Wertheimer's[1] findings
which are concerned with the effectiveness of particular and
forcible impressions of form.]

(2)

The objective is again out of reach on the other side of the
bars ; in the room itself, close to the bars, lies a piece of wire,
but wound in an oval coil and, therefore, too short to be used
as a stick ; a small box which has been used in other experiments has also been left. (16.3.1914) Sultan is brought
in, does not seem to see the wire, hesitates a moment, perplexed, and then breaks off a board from the lid of the box,
with which he immediately pulls the objective toward him.

[1] *Experimentelle Studien uber das Sehen von Bewegung, Zeitschrift für
Psych.*, vol. 61, p. 161 seqq. Cf. p. 243 seqq. In my subject, even
when I could least have expected it, I kept on discovering references
to this paper.

As this solution is already familiar, the board is removed, and a new objective is put there. Sultan looks round, does not seem to see the wire, and turns towards a piece of wire sticking out on the wall, pulls it off, and tries in vain to reach the objective—it is too short.

It now seems sufficiently clear that the animal has not seen the coil at all, which does not stand out against the sanded floor. Assistance, meant to arouse his attention, is given him by lifting the wire from the ground indifferently, and immediately putting it down again. Sultan looks round, at once picks it up, pulls at it impatiently and unmethodically with his teeth. A piece of wire comes undone ; the animal seizes it in his hands, bends it still straighter, and drags the objective towards him with the wire that is still half coil and only partly undone. The solution is undoubtedly *genuine*, but the way of treating the wire differs very much from a human adult's. To begin with, the coil is pulled blindly lengthways, without any regard to the way in which it is coiled ; then, when, by pulling with the teeth, a part of the coil (a loose end) comes unwound, it is stretched further in a perfectly sensible manner. But, while a man would not be satisfied without methodically unwinding the whole thing, Sultan evidently has no such considerations, but uses the wire as an implement at once.

The previous experiment with the gymnastic rope is repeated, and made more difficult. The objective is on the same spot (about two and a half metres' distance from the gymnastic apparatus), but the rope, instead of hanging loose, is laid in three firm coils starting from the hook, around the upper cross-beam, into which the hook is screwed. The coils are neat and orderly, do not cross each other, and can be easily surveyed by the human observer. The free end of the rope is now the part furthest away from the objective and hangs only about thirty centimetres down from the cross-

beam. (10.4.1914.) As soon as Chica sees the objective, she
climbs up the apparatus, seizes the middle coil of the rope
underneath the cross-beam, and pulls it once downwards,
then again with increased strength, so that the rope, except
for one coil nearest the hook, is thrown over the beam, and
hangs down. Without bothering about this last coil, she
now tries to swing herself at once to the objective ; twice
running she is unsuccessful, as the rope like this is too short
and cannot be swung properly.   Instead of remedying this
deficiency, Chica tries a third time, starting off with a still
bigger jump ; she jumps away from the rope in a big curve
through the air and towards the objective, seizes it and tears
it down with her as she falls. Apart from the gymnastic
feat, this proceeding acts as a translation from Sultan's
behaviour, above described, to the other kind of experiment.
The solution is genuine, and the energetic attempts to make
the rope hang down are made as soon as the animal has surveyed the situation ; but she takes no notice at all of the
nature of the coils, she merely seizes the rope in the middle
and pulls it down. The fact that, in spite of the disturbing
effect of the last coil, it yet remains unnoticed, makes the
analogy perfect. Although it leads to the solution, the
method at first looks untidy and unordered.

Sultan is tested on the same day. From the beginning
he seems weary and lazy, does not trouble much about the
objective, but after one attempt to use sticks for beating,
he climbs the apparatus and lazily unwinds one coil. Obviously he is not concentrating, and soon after he runs away
to play. When he comes back after a while, he again seizes
the rope—*but by the hanging end, so that, when jerked, the
coils become still tighter*—and pulls apathetically. After
twenty minutes the experiment is broken off, undecided.
The animal is too drowsy, and when all his movements are so
apathetic, it is not clear in how far they are connected, and

what the different phases of his conduct mean.[1] But, considering Sultan's manner of unwinding the coil of wire, there is a possibility that he meant to pull down the rope, but the second time attacked the thing blindly and without any regard for its structure ; indeed, the actual proceeding, objectively considered, was in the very opposite direction.

The assiduous Rana gives no cause for doubts such as these. (23.4) The rope is wound (in somewhat firmer and tighter coils) *four* times round the cross-beam, the separate coils neither crossing nor touching each other. Rana espies the objective, at once climbs the apparatus, hangs by her hands from the cross-beam, where the rope usually hangs, and quite unmistakably indicates the motion of swinging herself towards the objective. Immediately thereupon she begins to pull at the rope in a downward direction, but she seizes it blindly, catches the *uppermost part*, which is laid over the bar in the shape of a coil, and, when it does not give way to her pulling, but only slips a little to and fro in the hook, *makes obvious efforts to slip off the knot from the hook.* In this she is not successful, so she turns to the free end, throws it round properly once, then after an interval again, and soon has the whole rope hanging free. She tries to swing herself to the objective at once, but several times fails to reach it, as the distance is too great, so once more she sets herself to the manipulation of the implement. The only thing that can still be altered is the fixing of the knot, and *actually Rana does not stop until the knot is off the hook.* Somewhat puzzled when she finds the rope free in her hand, she takes it up to the cross-beam, and slowly winds it round her neck.

In so far as Chica and Rana try in this experiment to put

---

[1] With reference to previous explanations (see above, p. 16) we must conclude that experiments can only be carried out on animals when they are in a fresh condition ; this is obvious, in any case.

the rope into place for their purpose, the experiment has a positive result. At the same time, however, we discover that the really critical part of the task does not lie in that operation but in mastering the structural connexion of the rope and the beam. Chica may be too impatient to unwind the rope properly, but when we see what Rana does with the rope at the beginning and toward the end of the experiment, a second possibility is suggested : Rana does not treat the rope as if *she could visually apprehend the course of the windings* —as an adult could—*but regards it as we would a tangle of string*. We, when we do not see the ends of such a tangle immediately, and are too heated to trace them out separately, will seize the whole thing angrily and tear at it to unravel it ; so exactly does it look when Rana puts her hand into the middle of the rope-coils, and Chica too, excepting that she is, in addition, careless. It, therefore, seems possible that this relatively simple arrangement is at the very start confused ¨ to a chimpanzee, and as *optically incomprehensible* as much more complicated combinations are to us —such as tangled thread or wire or, to the author, even folding chairs. This is not contradicted by the fact that Rana got the rope to hang in a relatively short time, for her movements when doing this were not by any means completely sure, did not look as if they corresponded " to a plan " based on the arrangement of coils, and, indeed, one felt that success was partly a matter of chance. The fact that Sultan, even though drowsy, pulls at the rope in the direction *opposite* to that required, confirms our suspicion that chimpanzees do not grasp complicated forms as well as men.[1] I take it for granted that it is known that children up to the fourth year of their life, and even later, *treat* and, perhaps, actually

---

[1] Of course it remains true nevertheless, that the animals distinguish such different shapes as sticks, hat-brims, and shoes with complete assurance (cf. above, p. 36).

see coiled rope in the same way as it is thought the chimpanzee does ; but probably there will be individual differences even among adults.

This is not yet all : Rana, not reaching the objective with the rope stretched out as far as it will go, busies herself with the knot above, and this activity seems to arise directly from the unsuccessful effort, thus becoming like an attempt to improve still further the lie of the rope. Thus we get the following possibility : Rana cannot distinguish between the *coils* wound round the beam, and the externally similar *knot*, by which the rope is attached to the hook, whose function, however, is quite different. This may also be expressed by saying that *Rana has no insight into the manner in which the rope is held*. Her first attempt to unhook the knot may be explained by the fact that the confusion of coils brought about an error which even a full-grown man might commit, provided it is sufficiently complicated. Afterwards, when the rope is hanging quite freely and Rana yet tries to get the knot to " drop," it seems very likely that she does not understand (and certainly that she has not noticed) how the rope is fastened. It should be recalled that even in the simplest attempts to create implements (pulling the objective tied to a string) there was a question whether the chimpanzee ever sees more in a connexion than rough optical contact (see p. 27).

It is possible that both factors, the optical and the technical, are internally connected, as the simple technical device (the knot slung over the hook) represents structurally an optical task (Wertheimer) ; but the technical factor further bears on the question of how much the chimpanzee knows about the gravity and falling of objects. All this must be treated in greater detail in further experiments.

[It does not seem at all stupid when Rana, at the beginning, the rope being still coiled, hangs on to the cross-beam, and indicates that she is going to swing herself towards the objective ; and it certainly does not look as if the animal really were going to attain her goal in this way. It brings to mind rather that performance described before (p. 64) when Rana wanted to get the door of a cage free and, when unsuccessful, walked through the door of another cage close by, " as if buried in thought ". In that case it did not look either as if she expected to find the objective in the box, or to get nearer to it thus. The impression made on the observer by this proceeding could perhaps be best described as a sort of " expressive gesture ", expressing some such state as : " It all depends on getting through the door—to swing there from here, that is what matters ! " (The animal is entirely deprived of expression in speech ; *we* mutter words of this kind to ourselves, even when no one can hear us, e.g. when speaking is of no use, but we are " so full of the thing " that our tongue starts to utter words. Rana's limbs are her tongue.)]

The experiment to be described was only made when a similar, but harder, test had turned out decidedly badly. (29.3) The iron hook is unbent, so that the rope-knot can be unhooked or hooked with equal ease ; the rope is taken down and put on the floor in a few coils exactly under the hook ; the objective hangs as in the original test (p. 57). When Sultan is brought in, he tries one after the other of his usual methods ; he fetches sticks, boxes, and pulls the keeper and the observer under the objective ; we allow none of these things ; sometimes he seizes the rope, he may even lift it a little, but none of his movements indicate the solution of " hanging it up " ; it rather looks as if he were about to beat at the objective with the rope and is only prevented from so doing by its height.

Chica is admitted. She actually picks up the rope, even

drags it after her on to the cross-beam, pays not the slightest attention either to the hook or to the cross-beam, from which alone the rope could be hung, but makes movements as if to beat about for the objective. Finally she hangs on to a trapeze, which is attached to the same apparatus, but further to the side, jumps hard sideways towards the objective, lets go, springs through the air, and pulls the objective down with her as she lands. Our ordinary human ideas of gymnastics are not sufficient to meet Chica's case when preparing experiments for her.

In the afternoon care was taken that the trapeze could not be used; but although both animals lifted the rope again several times, neither Chica nor Sultan made any attempt to hang it up; and Chica, at the end, sprang without any implement from the cross-bar in a wide parabola to the objective, which she actually succeeded in pulling down with her.

Supplement 1916. This experiment, performed on Sultan, Grande, Chica, and Rana, still gives the same negative result; the unwinding of the rope, on the other hand (the experiment had not been repeated in the interval), had quite a different issue with Chica, and a somewhat different one with Rana (8.3.1916). Chica perceives the objective, climbs up to the cross-beam at once, unwinds the rope as methodically as a human being would, and then swings herself over to the objective. Rana does not proceed quite so definitely; though also with greater assurance than before. The difference may quite easily be that the animals are now older, which may have heightened their power of attention, or have further developed their visual functions; or they may have been helped by frequently playing with the rope in the interval, though, it must be added, the animals were not let in to the place where the rope hung during the six months previous to this after-experiment. Thus *the low degree* of optical

apprehension[1] assumed in the above discussions is not necessarily characteristic of chimpanzees, for a certain improvement is possible with them, just as with human children,
though in quite a different measure. But unfortunately,
the other experiments showing how the animals deal with
such, or somewhat more complex, forms, do not allow any
change in the opinions I formed two years ago ; a small
difference of degree is not of great moment and, therefore, I
leave the statements I made following on the first impressions,
which remain substantially correct.

(3)

The objective is hung at a spot high up ; a few metres
away is a box, which is open at the side so that one can see
three heavy stones in it (15.4.1914). Sultan comes up on the
closed side of the box and immediately starts pulling it towards
the objective. As it scarcely moves from the spot, he lets
go, looks into it, and carefully takes out one cf the stones.
Then he again begins to pull with great effort, gives it up,
and removes the second stone from the box. Without taking
any notice of the third, he pulls and pulls until he gets the
box underneath the objective. The experiment is imme-
diatly repeated ; Sultan pulls first at the box, then takes out
one stone, and pulls the two others with the box underneath
the objective, although this costs him a stiff effort. A third
experiment has exactly the same result as the first one. At
the fourth trial, Sultan pulls for a moment, then unpacks

---

[1] Motor-factors may play some rôle ; but in such cases they are applied
with too great facility, when it is a question of theory ; and the *nature* of
these factors, as well as their connexion with optics, is sometimes treated
of in a way which cannot be considered a model of empirical proceeding.
All the more care must be taken that such theories are not treated as
proven facts.

the three stones one after the other all at once (16.4). Four stones are in the box ; Sultan pulls a little at it, then rolls the four stones one after the other out of the box by a relatively great effort, and attains the objective with the empty box.

A month later (29.5), in other surroundings, the objective is again hung high up and, this time, the box is filled with sand up to the very top (which is open), and placed a long way off. Sultan goes immediately towards the box, dives into it with both his hands, and eagerly shovels out the sand. After a while, when there is still a great deal of sand left, he begins pulling at it again, on this occasion throwing the box on its side (very likely accidentally), so that more of its contents fall out, but he still cannot get away from the spot, as the rest of the sand is a heavy weight. He goes on unloading with both hands, eventually pulling the box underneath the objective, but without emptying it altogether, and, consequently, at some trouble.

One must not think that the chimpanzee, every time he sees stones in a box, begins by taking them out : Sultan takes out the stones every time, *when his pulling at the box has proved to be in vain*, and he also shows a strong tendency to limit this " auxiliary action " to the strict minimum. Here again, an experiment with less gifted animals will give us the information we want.

(18.4) Chica seizes the box, containing three stones, and pulls it (without any investigation as to the unusual weight) with all her might underneath the objective. On repeating the experiment, the weight is increased by a heavy stone which is put into the box *before Chica's eyes*. She pulls and pulls, without being able to budge the box, and finally gives up the useless effort, without having even touched the stones.[1]

Instead of the box, this time the *ladder* is to be used, which is lying on the floor some distance away, loaded with six

[1] In the autumn of 1914 a similar problem was solved fairly well.

heavy lava blocks. (14.5) Grande tries with a tremendous effort to pull the ladder underneath the objective. Not successful, she drags away one, then another of the blocks, and uses them as a *substitute for the box*. She is not successful, so she seizes the ladder again, drags it along the ground *with the four remaining stones lying on top*, quite close to the objective, and stands it up in that curious manner, which will be described later ; only now, all the stones fall off. The first two stones were certainly not taken away from the ladder to make it lighter ; they were meant, from the start, to be used as building material ; the acts of taking them down and dragging them from the ladder to the objective are merely parts of *one* action. Neither during the pulling along nor the setting up of the ladder does any movement occur that is intended to get rid of the remaining stones, which fall finally, only by accident, when the ladder is lifted.

An experiment with the box filled with stones worked out in the same way. (15.7) Grande tries first to break off a stick from the partition. When this fails, she approaches the box, but does not pull ; she takes out a stone, carries it underneath the objective, puts it carefully in a perpendicular position, glances upwards, does not get on top (it is too low), returns to the box, and with a really great effort pulls it towards the objective. On the way she stops a little, lifts up one of the stones, but leaves it, and, by an extraordinary effort, pulls the box right under the objective. In this proceeding the only suspicious factor is that, in the beginning, the box is not considered as a tool at all, as it is immediately in all other cases. From similar experiments with Grande, it would seem quite possible that she sees the box is very heavy, without her consequently making an attempt to lighten it. The later fact of lifting the stone looked as if she was going to use it as building material.

Rana (15.4) begins to drag the box towards the objective,

without appearing to notice the great weight ; on turning, the stones tumble out, and Rana is startled ; she evidently does not expect such a thing. After that she does not go near the box again, but tries eventually to pull the observer underneath the objective.

This experiment with Rana was the first of the whole group : I considered the task an easy one and only wanted to verify my opinion by testing the least gifted animal. The result is curious enough and recalls very clearly the obstacle experiments (compare above, p. 59 seqq.) ; the internal relation between the two problems is obvious (here, as there, an otherwise quite indifferent body is by its mere presence on the spot, in the way).

Even if Sultan does *not* begin by taking out *all* the stones one by one, that does not make the solution any less important ; it is merely æsthetically faulty in the eyes of the educated European ; the animal made the same æsthetic fault during the uncoiling of the wire as did Chica when unwinding the gymnasium rope.

(4)

When the chimpanzee is not very excited or careless, he will usually, as previously reported,[1] not use implements (sticks and boxes) which are not big enough for his purpose. Often, it is true, he brings them along, but as soon as the tool is close to the critical distance, his formerly active movements begin to falter, something has happened that had not happened further away, and, whatever may be the nature of this occurrence, it has the effect of putting a feeble end

---

[1] Exceptions are already known. Compare also what follows. Of course the chimpanzee does make an attempt, when the implement is *nearly* long enough.

to the energetic action. When the chimpanzee's tool is only a small stick, he goes only just up to the critical distance, and pokes once, or throws his stick in the direction of the goal; but an observer with any experience will have been able to indicate at once the moment when the fresh hue of determination faded; the rest, then, is not a practical endeavour, but merely the expression of discouraged desire.

Besides this paralyzing effect, an increase in the critical distance may influence the animal positively, too; but still not in the direction of practical progress. But if the result represents an error on the part of the animal, it is, at any rate, a " good error ".

It must be said that at the beginning happenings seem very curious: Chica is put a second time under observation while trying to reach an objective with the stick (26.i.1914). As she cannot quite reach it, she seizes a second stick which is a little shorter even than the first, puts it with its flat side against the equally flat side of the first stick, puts her hand round both of them carefully, and keeps fishing thus for the objective, although she does not extend the tool thus, or in any other way obtain a practical result by adding the second stick. Owing to the way in which she holds the short stick pressed against the long one, the short one does not even touch the ground. It might, of course, be said that the animal is too stupid to realize the senselessness of her proceeding. Roughly this may be true, *once the procedure has arisen*. The psychologist will, however, ask with astonishment how an animal, which, the day before, could barely manage to handle the stick with all the awkwardness of the beginner (though obtaining in a few days as much practice as possible) achieves all of a sudden this striking performance. What *started it* is a puzzle, especially as Chica, in the middle of her vain efforts, turns to the second stick and presses it so carefully against the other, that the whole action doubtless presents an attempt at tool-improvement.

Similar things can be seen often, but only so long as the use of the stick is not very familiar. A short time ago, Tschego, who experimented but little, on seeing that she could not reach the objective with her blanket, placed a stick on it, grasped the cover so that her fingers held the stick at the same time, and continued her useless efforts in this way; showing that, in this instance too, the improvement of the tool was the evident aim.

Rana, who, so to speak, can keep nothing in her brain, reaches one further stage in the experiment in which she had to jump with the stick; a stage which, in other animals, perhaps does not become visible. Curiously enough, she cannot manage to *strike at* an objective hung high up. Even after years (1916) of practice at using the stick as an extension of the arm, she is still quite clumsy, not knowing how to take hold of it correctly. She may raise the stick, but the next moment will use it again as a jumping-stick. Thus it happens that short pieces of wood, which might be used for beating down the objective, but would be of no use as vaulting-poles, if not actually utilized as poles (which is impossible), *seem* again and again to be going to be put to this use. Over and over again she puts small sticks of about thirty centimetres on the floor, raises one foot, as if to climb, and puts it down again (see p. 73). So too, once, with a box experiment. (15.4) Rana has placed a box underneath the objective, but, unable to reach it, she fetches a tiny little stick of about forty centimetres, puts it on the box in a position for a jump, and repeatedly makes the bodily movements preparatory to a jump, although, to keep the bottom end of the stick on the box at all, she has to stand quite bent—an attitude, of course, impossible for jumping. After a while, she fetches some more little sticks, holds them one against the other in her hand, but, of course, does not jump. Suddenly she changes her tactics, keeps only two sticks out of the bunch, and *puts them carefully end to end so that they look to the eye like a stick of twice the length* ; the ends of the two sticks are next to each other, though a distance of two fingers' width only, and so have to be held together with the hand. The whole thing is again set up like a jumping-stick, and the foot is raised as if to climb it. As Rana likes to attempt over and over again things which are in practice impossible, there is plenty of time to study her actions exactly. There is certainly no question of accident, for if the sticks slip and come together, they are always put back into a position which makes them at least look like one long stick, while they are held with the hand. It is astonishing to note how, apparently, the " optics " of the situation is decisive for the animal, how the endeavour to solve the problem takes no account of the " technically physical " point of view, but considers solely the optical aspect. The two sticks must be held together by the hand, and so, what according to the optical impression constitutes a solution by improvement of the implement, actually remains valueless. I may add that Rana really tries to utilize this lengthened stick.

Are the two sticks ever combined so as to become technically useful ? This time Sultan is the subject of experiment (20.4). His sticks are two hollow, but firm, bamboo rods, such as the animals often use for pulling along fruit. The one is so much smaller than the other, that it can be pushed in at either end of the other quite easily. Beyond the bars lies the objective, just so far away that the animal cannot reach it with either rod. They are about the same length. Nevertheless, he takes great pains to try to reach it with one stick or the other, even pushing his right shoulder through the bars.[1] When

[1] This is not in contradiction to the statement made above ; in order not to discourage the animal from the very beginning, I put the objective only just out of reach of the single stick.

everything proves futile, Sultan commits a "bad error", or, more clearly, a great stupidity, such as he made sometimes on other occasions. He pulls a box from the back of the room towards the bars ; true, he pushes it away again at once as it is useless, or rather, actually in the way. Immediately afterwards, he does something which, although practically useless, must be counted among the "good errors " : he pushes one of the sticks out as far as it will go, then takes the second, and with it pokes the first one cautiously towards the objective, pushing it carefully from the nearer end and thus slowly urging it towards the fruit. This does not always succeed, but if he has got pretty close in this way, he takes even greater precaution ; he pushes very gently, watches the movements of the stick that is lying on the ground, and actually touches the objective with its tip. Thus, all of a sudden, for the first time, the contact " animal-objective " has been established, and Sultan visibly feels (we humans can sympathize) a certain satisfaction in having even so much power over the fruit that he can touch and slightly move it by pushing the stick. The proceeding is repeated ; when the animal has pushed the stick on the ground so far out that he cannot possibly get it back by himself,[1] it is given back to him. But although, in trying to steer it cautiously, he puts the stick in his hand exactly to the cut (i.e. the opening) of the stick on the ground, and although one might think that doing so would suggest the possibility of pushing one stick into the other, there is no indication whatever of such a practically valuable solution. Finally, the observer gives the animal some help by putting one finger into the opening of the stick under the animal's nose (without pointing to the other stick at all). This has no effect ; Sultan, as before, pushes one stick with the other towards the objective, and as this pseudo-solution does not satisfy him any longer, he

[1] The way in which he does that is reported on p. 174.

abandons his efforts altogether, and does not even pick up
the sticks when they are both again thrown through the
bars to him. The experiment has lasted over an hour, and
is stopped for the present, as it seems hopeless, carried out like
this. As we intend to take it up again after a while, Sultan
is left in possession of his sticks ; the keeper is left there to
watch him.

Keeper's report : " Sultan first of all squats indifferently
on the box, which has been left standing a little back from
the railings ; then he gets up, picks up the two sticks, sits
down again on the box and plays carelessly with them. While
doing this, it happens that he finds himself holding one rod
in either hand in such a way that they lie in a straight line ;
he pushes the thinner one a little way into the opening of the
thicker, jumps up and is already on the run towards the
railings, to which he has up to now half turned his back, and
begins to draw a banana towards him with the double stick.
I call the master : meanwhile, one of the animal's rods has
fallen out of the other, as he has pushed one of them only a
little way into the other ; whereupon he connects them
again ".[1]

The keeper's report covers a period of scarcely five minutes,
which had elapsed since stopping the experiment. Called
by the man, I continued observation myself : Sultan is
squatting at the bars, holding out one stick, and, at its end,
a second bigger one, which is on the point of falling off. It
does fall. Sultan pulls it to him and forthwith, with the

[1] The keeper's tale seems acceptable to me, especially as, upon
inquiries, he emphasized the fact that Sultan had first of all connected
the sticks in play and without considering the objective (his task).
The animals are constantly poking about with straws and small sticks
in holes and cracks in their play, so that it would be more astonishing
if Sultan had never done this, while playing about with the two sticks.
There need be no suspicion that the keeper quickly " trained the
animal " ; the man would never dare it. If anybody continues to
doubt, even that does not matter, for Sultan continually not only
performs this act but shows that he realizes its meaning.

greatest assurance, pushes the thinner one in again, so that it is firmly wedged, and fetches a fruit with the lengthened implement. But the bigger tube selected is a little too big, and so it slips from the end of the thinner one several times ; each time Sultan rejoins the tubes immediately by holding the bigger one towards himself in the left and the thinner one in his right hand and a little backwards, and then sliding one into the other.[1] (Plate III.) The proceeding seems to please him immensely ; he is very lively, pulls all the fruit, one after the other, towards the railings, without taking time to eat it, and when I disconnect the double-stick he puts it together again at once, and draws any distant objects whatever to the bars.

The next day the test is repeated ; Sultan begins with the proceeding which is in practice useless, but after he has pushed one of the tubes forward with the other for a few seconds, he again takes up both, quickly puts one into the other, and attains his objective with the double stick.

(1.5) The objective lies in front of the railings, still farther away ; Sultan has three tubes to resort to, the two bigger ones fitting over either end of the third. He tries to reach his objective with two tubes, as before ; as the outer one keeps falling off, he takes distinct pains to push the thinner stick farther into the bigger one. Contrary to expectations, he actually attains his objective with the double tube, and pulls it to him. The long tool sometimes gets into his way when doing this, by its farther end getting caught between the railings, when being moved obliquely, so the animal quickly separates it into its parts, and finishes the task with one tube only. From now on, he does this every time when the objective is so close that *one* stick is sufficient, and the

---

[1] The illustration is culled from a cinematograph film that was taken a month later in other surroundings, more suitable for photographs.

PLATE III.   SULTAN MAKING A DOUBLE-STICK

double-stick awkward. The new objective is placed still farther away. In consequence, Sultan tries which of the bigger tubes is more useful when joined to the thin one; for they do not differ very much in length (64 and 70 cms.), and, of course, the animal does not lay them together in order to compare their lengths. *Sultan never tries to join the two bigger tubes*; once he puts them opposite to each other for a moment, not touching, and looks at the two openings, but puts one aside directly (without trying it) and picks up the third thinner one; the two wide tubes have openings of the same size.[1] The solution follows quite suddenly: Sultan fishes with a double-stick, consisting of the thinner one and one of the bigger ones, holding, as usual, the end of the smaller one in his hand. All of a sudden he pulls the double-stick in, turns it round, so that the thin end is before his eyes and the other towering up in the air behind him, seizes the third tube with his left hand, and introduces the tip of the double-stick into its opening. With the triple pole he reaches the objective easily; and when the long implement proves a hindrance in pulling the objective to him, it is disconnected as before.

According to observations in this experiment, Sultan never attempted to join tubes which would not have fitted together.[2] Once, when an experiment was to be shown to visitors, I put down the objective outside, and at the same time threw two different-sized tubes, which happened to be at hand, through the bars to Sultan. He took hold of them at once, the bigger one, as usual, in his left hand, the thinner in his right, and was already lifting his right hand to connect

[1] It can be shown that when the chimpanzee connects the double-stick he is guided by the relation between the two thicknesses of the tubes (compare *Nachweis einfacher Strukturfunktionen, etc. Abh. d. Preuss. Akad. d. Wiss.* 1918, *Phys.-Math. Kl.*, No. 2, p. 56 seqq.).

[2] In those cases in which mere observation does not lead to a definite conclusion, a trial is, of course, made. Compare experiment of 6.8.

the tubes, when he suddenly stopped, without carrying out his intention, turned the thick tube round, looked at its other end, and immediately dropped both tubes to the ground. I let him pass them out to me and discovered that both ends of the wider one happened to have a nodule, thus closing the opening; in these circumstances Sultan did not even attempt to connect the tubes. When I cut away this closed part, he made the trial at once.

(6.8) The wide tube is closed at one end; a wooden block is put into the other end before the experiment; it sticks out just a little, being somewhat narrower than the tube, so that a space is left between it and the tube. Sultan seizes the tubes, looks for a moment at the block in the hole, tries to squeeze the thinner tube into the narrow opening between the block and the side of the tube, fails, and straightway pulls out the stopper, throws it aside, and connects the tubes.

Sometimes, however, he experiences a difficulty, where one would least expect it. Holding both tubes in his hand and wanting to proceed as usual to connect them, he hesitates for a few moments and seems strangely uncertain; this is when the tubes lie in his hand in certain positions, namely, almost parallel, or else across each other in the shape of a very narrow " X ". This difficulty has now almost disappeared, but at first it occurred frequently. When Chica had, later on, adopted the same procedure, she showed exactly the same embarrassment when the two tubes were in this position, and her embarrassment was even more striking than Sultan's. As soon as the animals have again optically separated one tube from the other, the action proceeds quite smoothly. The optical factor in the situation, which at other times is a sure guide to the chimpanzee, making his acts and his behaviour appear the direct result of it, must in this case be so changed that it does not determine the motor factor quite so definitely. We humans can always see the relative positions of two tubes like this clearly enough not to be thus embarrassed, but if slightly complicated conditions (unfolding a folding-couch) be introduced, we too take some seconds to readjust, before our vision can dictate our movements.

In cases of pure *alexia* (Wertheimer) this uncertainty seems to be greatly increased. Gradually it becomes obvious that to understand the capacities and mistakes of chimpanzees in visually given situations is quite impossible without a theory of visual functions, especially of shapes in space.

In another experiment, further manufacture of implements is demanded of Sultan. (17.6) Besides a tube with a large opening, he has at his disposal a narrow wooden board, just too broad to fit into the opening. Sultan takes the board and tries to put it into the tube. This is not a mistake ; the different *shapes* of the board and the tube would tempt even a human being to try it, because the difference in thickness of both these objects is not obvious at first sight. When he is not successful, he bites the end of the tube and breaks off a long splinter from its side, obviously because the side of the tube was in the way of the wood (" good error "). But as soon as he has his splinter, he tries to introduce it into the still intact end of the tube ; a surprising turn, which should lead to the solution, were not the splinter a little too big. Sultan seizes the board once more, but now works at it with his teeth, and correctly too, from both edges at one end towards the middle, so that the board becomes narrower. When he has chewed off some of the (very hard) wood, he tests whether the board now fits into the sound opening of the tube, and continues working thus (here one must speak of real " work ") until the wood goes about two centimetres deep into the tube. Now he wishes to fetch the objective with his implement, but two centimetres is not deep enough, and the tube falls off the top of the wood over and over again. By this time Sultan is plainly tired of biting at the wood; he prefers to sharpen the wooden splinter at one end and actually succeeds so far as to get it to stick firmly in the sound end of the tube, thus making the double stick ready for use. In connexion with this treatment of the wood it must be remarked that, contrary to my expectation, Sultan bit away wood almost exclusively from *one* end of the board, and, even if he took the other end between his teeth for a moment, he never gnawed blindly first at one, and then at the other. His way of dealing with the tube was also very satisfactory. The one opening of the tube that had been spoiled by breaking its side is thereafter left unheeded. I had some anxiety for the other opening during the further experiment, but although Sultan, when the wood and splinter did not 'fit in, put his teeth into it several times, he never really bit into the side of the tube, so that the opening could still be used. I could not guarantee that each repetition of the experiment would turn out so well. Sultan evidently had a specially bright day.

The apes have often sharpened wood, moreover, before this experiment. For instance, if Grande wants to poke somebody through the bars, she will swiftly bite a board in two and thus get the splinters she needs. Sultan too, if there is no key about, will occasionally sharpen a piece of wood in order to poke about in the keyhole; a fact noticed over and over again in the literature of this subject. But I was never quite clear about this sharpening business, and, therefore, we now investigated whether Sultan would proceed rationally with the very hard wood, which he would never have separated into serviceable splinters in mere play or by accident, but which he would have to work at somewhat methodically.

It will be obvious, after all the foregoing, that the double stick is made as promptly, when the objective is too high up to be knocked down with *one* stick only, and also that Chica, having adopted the new method, will apply it on occasion to the jumping stick procedure.

V

THE MAKING OF IMPLEMENTS—(cont.)

BUILDING

WHEN a chimpanzee cannot reach an objective hung high up with *one* box, there is a possibility that he will pile two or more boxes on top of one another and reach it in that way. Whether he *actually* does this seems a simple question that can soon be decided. But if experiments are made, it is quickly seen that the problem for the chimpanzee falls into two very distinct parts : one of which he can settle with ease, whilst the other presents considerable difficulties. We think the first is the *whole* problem ; where the animal's difficulties begin, we do not, at first, see any problem at all. If in the description this curious fact is to be emphasized as much as it impressed itself on the observer, the report of the experiment should be divided into two parts in accordance with this fact. I shall begin with the answer to the question that seems to be the only one.

In one of the experiments described previously (p. 46), Sultan came very near putting one box on the top of another, when he found one insufficient ; but instead of placing the second box, which he had already lifted, upon the first, he made uncertain movements with it in the air around and above the other ; then other methods replaced these confused movements. The test is repeated (8.2) ; the objective is placed very high up, the two boxes are not very far away from each other and about four metres away from the objective ; all other means of reaching it have been taken away. Sultan drags the bigger of the two boxes towards the objective, puts it just underneath, gets up on it, and looking upwards, makes ready to jump, but does not jump ; gets down, seizes the other box, and, pulling it behind him, gallops about the room, making his usual noise, kicking against the walls and showing his uneasiness in every other possible way.[1] He certainly did not seize the second box to put it on the first ; it merely helps him to give vent to his temper. But all of a sudden his behaviour changes completely ; he stops making a noise, pulls his box from quite a distance right up to the other one, and stands it upright on it. He mounts the somewhat shaky construction, several times gets ready to jump but again does not jump ; the objective is still too high for this bad jumper. But he has achieved the essential part of his task.

(12.2) Some days previously Chica and Grande learnt from Sultan and myself how to use *one* box ; they do not yet know how to work with *two*. The situation is the same as in Sultan's experiment. Each of the animals forthwith seizes a box ; first Chica, then Grande, will stand under the objective with her box, but there is no sign of an attempt to put one on top of the other. On the other hand, they hardly get up on their own boxes ; though their feet are lifted, they put them down again as soon as they glance upwards. It is certainly not a matter of accident, but the result of that upward glance at the objective, when both Chica and Grande proceed to stand the box upright (compare Sultan, p. 46) ; a measurement of the distance with the eye

---

[1] All the animals showed a strong aversion to the room in which these experiments were carried out, not because of the experiments—those they did not mind—but because of the unbearable dry heat that existed there most of the time. In those days, for outside reasons, I could not make my experiments anywhere else, but later I avoided the room whenever possible. Some stupidities that were observed here were very likely, partly symptoms of fatigue.

leads to this change of plan ; it is a sudden and obvious attempt
to meet the needs of the situation. Finally, Grande seizes
her box and tears about the room with it, in a rage, as Sultan
did before. Just as he did, she calms down unexpectedly,
pulls her box close to the other one, after a glance at the
objective, lifts it with an effort, puts it clumsily on the lower
one, and quickly tries to get up on it ; but when the upper
box slips to one side during this operation, she makes no
move, and lets it fall altogether, quite discouraged. In
principle Grande solved the problem too, so the box is lifted
by the observer, placed firmly on the lower one, and held there,
while Grande climbs up and reaches the objective. But she
does all this with the greatest mistrust.

(22.2) Grande, Chica, and Rana are present. Grande
carries first one, then the other box underneath the objective,
but handles them in such a way as to create the impression
that she is perplexed ; she does not put one box on the other.
This looks very like the condition of " lack of direction "
which sometimes influenced Sultan and Chica when dealing
with the two bamboo sticks. Suddenly Chica springs up
beside Grande, puts one box on the other without further
delay, and gets up on top. It is hard to say whether this
was an after-effect of the previous attempt and Grande's
example, or an independent solution, helped perhaps by
Grande's " messing round ".

A new objective is hung up ; Rana now puts one of the
boxes flat underneath the objective and the second one immediately on top of it (also flat) ; but the arrangement is
too low, and the animals prevent each other from improving
it, as they now all want to build on their own, and at the same
time. Knowing Rana, I am inclined to assume that this
is a case of imitation of what she has just seen, or, at any rate,
what she saw was of great help to her ; but this question is
not important here.

K

A number of further experiments, which, however, did not quickly lead to greater assurance, as in other cases, will be described later. After the animals had become accustomed to putting one box on another as soon as the situation called for it, the question arose as to whether they would make further progress in the same direction.

The tests (higher objective, three boxes at some distance) first resulted in Sultan carrying out more difficult constructions, with two boxes on top of each other, perpendicularly so that they looked like columns, and, of course, enabled him to reach very high (8.4) ; he took the third box, to being with, to the place of construction, but left it standing beside him without using it, as he could then reach the objective by means of his column without it.

(9.4) The objective hangs still higher up ; Sultan has fasted all the forenoon and, therefore, goes at his task with great zeal. He lays the heavy box flat underneath the objective, puts the second one upright upon it, and, standing on the top, tries to seize the objective. As he does not reach it, he looks down and round about, and his glance is caught by the third box, which may have seemed useless to him at first, because of its smallness. He climbs down very carefully, seizes the box, climbs up with it, and completes the construction.

Grande in particular progressed with time. Of the smaller animals she was the strongest and by far the most patient. She would not allow herself to be diverted by any number of mishaps, the collapse of the structure, or any other difficulties (partly created involuntarily by herself), and soon was able to put three boxes on top of each other, like Sultan (see Plate IV). She even managed once (30.7. 1914) a beautiful construction of four boxes, when she found a fairly big cage near by, whose flat surface allowed of the addition of the three remaining parts with safety. When, in the spring

PLATE IV. GRANDE ON AN INSECURE CONSTRUCTION (NOTE
SULTAN'S SYMPATHETIC LEFT HAND)

of 1916, an opportunity was again given for making higher constructions, Grande was still, even after this long interval, relatively the best of them all and quite as good an architect as before. High constructions composed of four objects gave her some difficulty, but with obstinate effort she managed them with considerable success (see Plate V).

Chica also builds towers composed of three boxes without too many mishaps, but has not become so expert as Grande, because, impatient and quick by nature, she prefers dangerous jumps (with or without a stick) from the floor or from some low structure to the slow process of building. And in these she is often successful, while Grande in her own way has still a good deal of hard work to get done.[1] Rana scarcely gets beyond two boxes. Whenever she has got so far, she stops, and either goes on endlessly trying out miniature vaulting-poles, or else (a frequent occurrence) she places the upper box open side up, and then carries out an irresistible impulse to sit down beside it ; once she is there, she feels too comfortable to get up again and to continue building. Konsul never built, Tercera and Tschego got no further than some feeble attempts, Nueva and Koko died before they could be experimented with.

Without doubt, constructions such as those achieved by Grande (see illustrations) are considerable feats, especially when one considers that the constructions of insects (ants, bees, spiders) and other vertebrates (birds, beavers), though they may, when finished, be more perfect, yet are built by a very different and much more primitive process, from an evolutionary point of view. The following accounts will show that the difference between the clever but clumsy constructions of a gifted chimpanzee, and the firm and objectively elegantly-spun web of a spider, for instance, is one of *genus*,

[1] Chica sometimes also uses the stick to beat with, instead of as a jumping-stick (cf. Plate VI).

which, of course, should be obvious from what has been already said. But, unfortunately, I have been asked by otherwise intelligent spectators of these constructions, "whether this is not instinct "? Therefore I feel obliged to emphasize the following particularly : the spider and similar artists achieve true wonders, but the main *special* conditions *for this particular work alone* are within them, long before the incentive to use them occurs. The chimpanzee is not simply provided for life with any special disposition which will help him to attain objects placed high up, by heaping up any building material, and yet he can accomplish this much by his own efforts, when circumstances require it, and when the material is available.

Adult human beings are inclined to overlook the chimpanzee's real difficulty in such construction, because they assume that adding a second piece of building material to the first is only a repetition of the placing of the first one on the ground (underneath the objective) ; that when the first box is standing on the ground, its surface is the same thing as a piece of level ground, and that, therefore, in the building-up process the only new factor is the actual lifting up. So the only questions seem to be, whether the animals proceed at all " tidily " in their work, whether they handle the boxes very clumsily, and so forth. I myself never expected to be faced, through my observations, by a wider, and very much more important, question. That another special difficulty exists, however, should become obvious from the further details of Sultan's first attempt at building. May I repeat : When Sultan for the first time fetches a second box and lifts it (28.1), he waves it about enigmatically above the first, and does not put it on the other. The second time (8.2, compare p. 135) he places it upright on the bottom one, seemingly without any hesitation, but the construction is still too low, as the objective has accidentally been hung too high up.

The experiment is continued at once, the objective hung about two metres to one side at a lower spot in the roof, and Sultan's construction is left in its old place. But Sultan's failure seems to have a disturbing after-effect ; for a long time he pays no attention at all to the boxes, quite contrary to other cases, where a new solution was found and usually repeated readily (though Koko once forgot a solution). It may well be that for the chimpanzee (as for man) the practical " success " of a method is more important as an estimate of its value, than is really justifiable. (This is judging *ex eventu* in the bad sense.)

[It occasionally happens that one will start working at a mathematical or physical problem with perfectly correct premises, and calculate or think up to a point where one gets lost. The whole proceeding is then rejected, and only later will one discover that the method was quite right, and that the difficulty was only a superficial one and could easily have been overcome. If, when the difficulty occurred, the logical relations had been the sole determinant, and if these had been carefully examined, the obstacle would at once be found unimportant. The less one takes into consideration all the relevant conditions, the more one will be embarrassed by an apparent failure. And so, it is not surprising that the chimpanzee, who does not grasp certain parts of the situation at all clearly, is influenced just as much by a mistake methodologically unimportant, as by an error in principle, and will then, because subsidiary circumstances spoiled the first attempt, give up the whole thing in despair. Grande furnishes a good example, when she suddenly puts a second box on the first one ; the solution is not only objectively good in principle, it also appears with the character of genuineness ; but as luck would have it, a corner of the top box is upon a board nailed across the surface of the lower one, so that when the animal begins to climb up, the box slips sideways, and then

Grande lets it fall altogether, and shows distinctly by her behaviour that, for her, the *whole method* is now completely spoiled. But such incidents can only happen under one condition, i.e. that one side of the situation is not clearly grasped, and so we arrive at our chief point : the experiment with two boxes involves conditions which the chimpanzee does not quite understand.]

Further on in the experiment a curious incident occurs : the animal reverts to older methods, wants to lead the keeper by his hand to the objective, is shaken off, attempts the same thing with me, and is again turned away. The keeper is then told that if Sultan tries to fetch him again, he is apparently to give in, but, as soon as the animal climbs on his shoulders, he is to kneel down very low. Soon this actually happens : Sultan climbs on to the man's shoulders, after he has dragged him underneath the objective, and the keeper quickly bends down. The animal gets off, complaining, takes hold of the keeper by his seat with both hands, and tries with all his might to push him up. A surprising way to try to improve the human implement !

When Sultan now takes no further notice of the box, since he once discovered the solution by himself, it seems justifiable to remove the cause of his failure. I put the boxes on top of each other for Sultan, underneath the objective, exactly as he had himself done the first time, and let him pull down the objective.

[As to Sultan's effort to push the keeper into an erect position, I should like at the very beginning to rebut the reproach of " misunderstanding ", of " reading into the animal " ; the procedure has merely *been described*, and there is no possibility at all of its being misunderstood. But lest suspicions should arise, this case being an isolated one (an unjustifiable suspicion in any case, considering that Sultan tries to utilize both the keeper and me, not once, but over

PLATE V.   GRANDE ACHIEVES A FOUR-STOREY STRUCTURE

BUILDING

and over again, as a footstool), I shall briefly add a description
of similar cases : (19.2) Sultan cannot solve a problem, in
which the objective is outside the bars beyond reach ; I am
near him inside. After vain attempts of all sorts, the animal
comes up to me, seizes me by the arm, pulls me towards the
bars, at the same time pulling my arm with all his might
down to himself, and then pushes it through the bars towards
the objective. As I do not seize it, he goes to the keeper,
and tries the same thing with him. Later (26.3) he repeats
this proceeding, with the only difference that he first has to
call me with plaintive pleading to the bars, as this time I
am standing outside. In this case, as in the first, I offered
so much resistance that the animal could barely overcome
it, and he did not release me until my hand was actually
on the objective ; but I did not do him the favour (in the
interests of future experiments) of bringing it in.—I must
mention further, that one hot day the animals had had to
wait longer than usual for their water " course ", so that
finally they simply grabbed hold of the keeper's hand, foot,
or knee, and pushed him with all their strength towards the
door, behind which the water-jug usually stood. This became
their custom for some time ; if the man tried to continue
feeding them on bananas, Chica would calmly snatch them
out of his hand, put them aside, and pull him towards the
door (Chica is always thirsty).—It would be erroneous to
consider the chimpanzee unenlightened and stupid in these
matters. I must add that the animals understand the human
body particularly easily in its local costume of shirt and
trousers without any coat. If anything puzzles them, they
will investigate it on occasion, and any large change in the
manner of dressing or appearance (e.g. a beard) will make
Grande and Chica undertake an immediate and very interested
examination.]

After the encouraging assistance to Sultan, the boxes are

again put aside. A new objective is hung in the same place on the roof. Sultan immediately builds up both boxes, *but at the place where the objective had been hung at the very beginning of the experiment and where his own first construction had stood.* In about a hundred cases of using boxes for building, this is the only one in which a stupidity *of this kind* was committed. Sultan is quite confused while doing this, and is probably quite exhausted, as the experiment has lasted over an hour in this hot place.[1] As Sultan keeps on pushing the boxes to and fro quite aimlessly, they are once more put on top of each other underneath the objective ; Sultan reaches it, and is allowed to go. Only on *one* occasion did I see him similarly confused and disturbed.

The next day (9.2) it is clear that a particular difficulty must lie in the problem itself. Sultan carries one box underneath the objective, but does not bring the second one ; finally it is built up for him and he attains the goal. The new one immediately replacing it (the construction was again destroyed) does not induce him to work at all ; he keeps on trying to use the observer as a footstool ; so once more the construction is made for him. Underneath the third objective Sultan places a box, pulls the other one up beside it, but stops at the critical moment, his behaviour betraying complete perplexity ; he keeps on looking up at the objective, and meanwhile fumbling about with the second box. Then, quite suddenly, he seizes it firmly, and with a decided movement places it on the first. His long uncertainty is in the sharpest contrast to this sudden solution.

Two days later the experiment is repeated ; the objective is again hung at a new spot. Sultan places a box a little aslant underneath the objective, brings the second one up,

---

[1] I only noticed later that I used to strain the animals a little too much during the first months ; only with time did I develop the slowness of procedure adequate to the apes and to the climate.

and has begun to lift it, when, all the while looking at the objective, he lets it drop again. After several other actions (climbing along the roof, pulling the observer up) he again starts to build ; he carefully stands the first box upright underneath the objective, and now takes great pains to get the second one on top of it ; in the turning and twisting, it gets stuck on the lower one, with its open side caught on one of the corners. Sultan gets up on it, and straightway tumbles with the whole thing to the floor. Quite exhausted, he remains lying in one corner of the room, and from here gazes at both box and objective. Only after a considerable time does he resume work ; he stands one box upright and tries to reach his goal thus ; jumps down, seizes the second, and finally, with tenacious zeal, succeeds in making it stand upright also, on the first one ; but it is pushed so far to one side that, at every attempt to climb up, it begins to topple. Only after a long attempt, during which the animal obviously acts quite blindly, letting everything depend on the success or failure of planless movements, the upper box attains a more secure position, and the objective is attained.

After this attempt Sultan always used the second box at once and, above all, was never uncertain as to where he had to put it.

The report shows that after the first independent solution, the boxes were arranged on top of each other for Sultan *four* times ; in the experiments on Grande, Chica, and Rana, I gave the same amount of assistance three times after the first solution, which proved a good incentive to the animals to go on with that method. If I had let them get very hungry and then put them time and again into the same situation, they would probably have developed their building process without this interference. But what seemed to me more important, after my first experiences (instead of trying to see whether the chimpanzees would keep on building without encouragement, proceeding to constructions of three and four parts), was to examine minutely their *method* of building ; that was why, after they had once solved the first essentials of the problem, I encouraged them as much as possible to continue.

If putting the second box on the first were nothing more than a repetition of the simple use of boxes (on the ground) on a higher level, one would expect—after the other experiences—that the solution once found would simply be repeated. To Sultan and Grande it is quite a matter of custom—in the days of these experiments—to attain objectives by means of *one* box, as the tests show ; but neither succeeded easily in reproducing his *building* methods, and one glance at the description of the experiments will show that the first failure (merely practical) is not alone to be blamed for it Neither is a quite external factor the chief cause : it is true that the boxes are heavy for the little animals, and there are moments in the course of the experiments when they simply cannot manage the weight. But one has only to see with what energy and success they generally carry and lift their burden, when they build at all, and how completely perplexed they may become, too, even when they have the second box high enough (from a human point of view) merely to let it sink on to the lower one, to realize that the animals do *not* omit further building merely on account of the physical effort. Rather, they will be a little clumsy to begin with. But too much stress must not be laid on this ; for probably the abandoning of the method after the first trial, is connected, internally, with their other strange behaviour, their sudden fits of perplexity before the two boxes ; and this behaviour has nothing to do with clumsiness. The animal does not behave then like somebody accomplishing a task clumsily, but like someone to whom the situation does not offer any definite lead toward a particular action.

PLATE VI.   CHICA BEATING DOWN HER OBJECTIVE WITH A POLE

This inhibition, perplexity, or whatever one likes to call it, which may befall the animals in their first attempts, when obviously the solution " put second box up " has already appeared and they are proceeding to carry it out, was observed three times in Sultan, twice in Grande, and clearest of all, later (in the spring of 1916) in the adult Tschego, on the occasion when she was to place one box on the other for the first time. I want to emphasize again that at first everything goes well ; as soon as the animals are quite familiar with the situation, and are convinced that they cannot attain the objective with *one* box, a moment arrives when the second box is suddenly " drawn into the task ". They then drag it up (Tschego) or carry it just to the first box and all of a sudden stop and hesitate. With uncertain movements they wave the second one to and fro over the first (unless they let it drop to the ground immediately, not knowing what to do with it, as Sultan once did) and if you did not know that the animals see perfectly well in the ordinary sense of the word, you might believe that you were watching extremely weak-sighted creatures, that cannot clearly see where the first box is standing. Especially does Tschego keep lifting the second box over the first and waving it about for some time, without either box touching the other for more than a few seconds. One cannot see this without saying to oneself : " Here are two problems ; the one (' put the second box up ') is not really a difficult task for the animals, provided they know the use to which a box can be put ; the other (' *add one box to the other, so that it stays there firmly, making the whole thing higher* ') *is extremely difficult*." For therein lies the one essential difference between using one box on the ground and adding a second to the first : In the *former* case, on the homogeneous and shapeless ground, which does not claim any special requirements, a compact form is simply put down or else it is just dragged along (till underneath the objective) without being taken off the ground at all. In the *latter* case a limited body of special shape is to be brought into contact with a similar one, in such a way that a particular result is obtained ; and this is where the chimpanzee seems to reach the limit of his capacity.

A glance in retrospect will show immediately that the experiments before described with *one* box only on level ground, get over this difficulty ; but they are misleading, and, therefore, cannot give an adequate idea of the chimpanzee. Either the little animal *pulls* his box almost underneath the objective, or he *rolls* it there. In neither case does it matter whether the box is some centimetres or even decimetres to the right, left, in front, or behind. The ground is the same level everywhere, and the objective, in spite of these small differences in position, is easily attainable[1] and, therefore, in the hands of the chimpanzee (who does not see any problem at all) the box automatically, with a few quick movements, reaches a position of equilibrium, in which it can be used. Quite different is the experiment with *two* boxes. Here the chimpanzee already meets a *static problem* which he must solve[2], since the first and second box do not solve it by themselves, as the first box and the level ground did.

[These observations and discussions lead to the conclusion that the chimpanzee will, without effort, place a small box on a very big box under the objective (the surface of the big box being both optically and physically more like the ground) ;

---

[1] Considerable mistakes of this nature, with reference to the objective, are easily and " genuinely " corrected (compare Koko) ; no factor of shape of any higher degree enters into consideration here, but simply " distance."

[2] Probably he seldom solves it " genuinely " ; but it seems remarkable to me that cases occur (as those described) where at least the problem as a problem has an effect upon the chimpanzee and keeps him perplexed, since the solution does not appear. He could, after all, just let the second box drop somehow on to the first, and need not hesitate uncertainty may also be a good sign, on occasion.

and, as a matter of fact, once, where a big cage formed the lower part of the structure, the second small box was immediately put firmly on top of it.]

There are two kinds of statics involving insight, just as there are (compare above, p. 75 seqq.) two ways of mastering lever action. The one kind, the physicist's (centre of gravity, movement of a force, etc.), does not come into question here any more than in those countless cases in which man "correctly" lays or stands some things on others. Unfortunately, psychology has not yet even begun to investigate the physics of ordinary men, which from a *purely biological standpoint*, is much more important than the science itself, as not only statics and the function of the lever, but also a great deal more of physics exist in two forms, and the non-scientific form constantly determines our whole behaviour.[1]

However the naïve statics of man may have arisen, even the most superficial observation will show that "gravity" on the one hand and visual forms in space on the other, play just as important a part in it as forces and distances, considered abstractly, in strictly physical statics. At least *one* of those "components" must be in a very undeveloped state in the chimpanzee ; for the total impression of all observations made repeatedly on the animals leads to the conclusion that *there is practically no statics to be noted in the chimpanzee*. Almost everything arising as "questions of statics" during building operations, he solves not with insight, but by trying around blindly. And there can be no more striking contrast than that between genuine solutions arrived at suddenly and in one sequence, and the blind groping about with one box on the other, which is the procedure of construction when no lucky chance (like those described above) brings box on box, surface to surface. "Bringing a second box

---

[1] With experts, of course, this is saturated in all stages by physical science in the *strict* sense.

above the first one " (or a third or fourth—not realized as numbers, but as " more " or " others ") no doubt still comes into the category of " genuine solutions ", but the expression " to set one on top of the other " should be used with great caution, when meant to denote what the chimpanzee really does. These words suggest our (not necessarily scientific) human statics, and the animal possesses extremely little of this.

One may observe very similar facts in the first years of childhood. Very young children also, in attempting to pile one thing on another, try, by holding, and sometimes pressing, one against the other, to fix them in different and often curious positions. It is quite obvious that they too lack that kind of statics. But while human children, when about three years old, begin to develop the elements of this naïve physics of equilibrium, the chimpanzee does not seem to make any essential progress in this direction, even when he has plenty of opportunity to practise. For, although his uncertainty in the sphere of spatial forms and gravity soon discourages him less than at first (when he gives up all effort, in face of the conglomeration of boxes), yet even after success has strengthened his confidence, his gaily-undertaken work remains as much " mere trying " as at the beginning : a turning, pulling, twisting, tipping of the upper box on the lower, so that the animals, especially Grande, arouse admiration by their patience. One must not think that such a construction, even of three boxes only, can be accomplished in a few seconds ; the more scope the boxes give for various accidents, the smaller they are, the more boards they are made of, the longer the animals will have to work, and it has happened that Grand kept on building up her structure for ten minutes at a time, then tumbling down with it, beginning again, and so on, until she was quite exhausted, and altogether unable to continue at all.

In the confusion of this method of construction some features are particularly characteristic. If the upper box is brought into a position in which it stands quite satisfactorily from a static point of view, but in which it may still wobble a little (this motion having no significance), it is often taken, or turned, out of this good position, if either hand or foot discover the oscillaiton ; for the optics of the position has here no further noticeable significance for the chimpanzee's control over the situation. If by chance, or in any other way, the upper box comes into any position where it does not for the moment wobble, the chimpanzee will certainly climb up, even though a mere touch or friction at some point has for the moment steadied the box ; really it may be quite unsteady and may fall over at once if weight is put on it. Thus Sultan, quite as a matter of course, once tried to climb up on the second box when it was precariously balanced on one corner only of the under one. Whether one box, for example, projects quite far out sideways from the rest of the structure or not, seems to be a matter of indifference to the chimpanzee —and sometimes the third box does not fall, only as long as the fourth and the animal remain on top of it and steady it by their weight. So one sees what happens when the chimpanzee deviates for the first time quite definitely from his optically-led treatment of the situation ; probably because it no longer serves to meet his needs. Structures grow under his hand, and often enough he can climb them, but they are structures which, according to the rules of statics, seem to us almost impossible. For all structures that *we* know (and are familiar with optically) are achieved by the apes by chance at best, and, as it were, by the " struggle for not wobbling." If Grande's first three-box construction is examined (see above, Plate IV, which, I hope, is clear enough to illustrate the point), it will be seen that it is scarcely capable of " life " ; it is not able to stand alone ; at the moment when the photograph was taken, it actually was no longer standing of its own equilibrium, but thanks only to the well-balanced weight of Grande herself, who, in her turn, is holding fast to the objective above, and who cannot take it off or let go without tumbling down with the whole structure.[1] Such occurrences are quite frequent, only that the constructions often look even more perilous ; usually the catastrophe happens before there has been a quiet moment long enough to take a photograph.

From this description it will be seen that the animals partly replace the missing (everyday) statics of human beings by a third kind—that of their own bodies, which is taken care of automatically by a special neuro-muscular machinery. In this respect, the chimpanzee, it seems to me, is even superior to man, and he obviously draws an advantage from this gift. When he is standing on a structure, the balance of which would fill an onlooker with fear, the first suspicious wobbling of the structure is counteracted in the most masterly fashion by an instantaneous altering of the balance of the body, by lifting his arms, bending his trunk, etc., so that the boxes, under the animal, to a certain extent, share the statics of his labyrinth and cerebellum.   It can be said that in a great number of these constructions the animal itself, with the delicately-balanced distribution of its weight, contributes a certain element, without which the structure would collapse. But this is chiefly a *physiological* achievement, in the narrower sense of the word ; there is no question of a real " solution ". I must express a warning against an explanation which is too easy and quite inadequate in face of the actual facts, which would consider the animals merely too untidy and careless to build anything more stable.   The animals' work may make this impression on a novice, but longer observation

---

[1] And immediately after the photograph was taken, the mishap occurred.

of the tireless energy which Grande displays—as much in pulling down well-built structures because one part wobbles, as in building up structures which do not statically balance—will convince anyone that the real explanation lies deeper, and that, at least, those animals up till now observed, are chiefly hindered by the limits of their " visual insight "[1].

If the animals cannot even intelligently combine the building materials into one whole, it is not astonishing if they often are unable to understand or deal with an existing structure ; for the corresponding (naïve) human faculties are missing, and can only be acquired with difficulty ; nor is there any question of mere haste or untidiness here.   So it will sometimes happen that Grande (and others too), while standing on one box, will try to lift another up in spite of its being open on one side, and a corner of the first box projecting into it. So, at least in part, Grande by her own weight, which rests on both boxes, hinders the lifting of the second box, but in spite of that she takes the greatest pains to drag it up, tearing and shaking it until, in a rage, she finally gives up trying to accomplish what she herself, without realizing it, is preventing. In the same way it may happen that Grande will be standing on a box supported at each end by two others, like pillars,[2] and that then one of the lower boxes will strike her as suitable for building purposes ; if she can, she calmly pulls this out

---

[1] Nueva dealt with shapes in space so much more sensibly than any of the others, that one was led to think that she might have built differently, had she ever got as far as building experiments. That there is an " *optical* weakness " must be taken for granted in any case, because even in the simple " gravitational physics ", " *gravity* " is for the most part determined optically.

[2] Such a thing can *only* come about by chance. Not once did the animals purposely pile up their boxes on the *bridge principle*, although in several experiments I tried to suggest such a proceeding to them, by hanging the objective high up, by placing heavy, solid pedestals at its right and left, and laying a stout board handy, so that they had only to place it across in order to reach the objective by standing on its centre. The board was always used (by Sultan and Chica) as a jumping-stick. Similarly all other experiments failed, in which the principle of using *two* forces at the same time plays any rôle.

at the side, and is very much startled when, together with the box on which she is standing, she tumbles to the ground (as must happen). I saw this even in 1916; there is simply no perceptible improvement.

[On the other hand the animals seem to learn that it is well not to turn the open side of a box upward in building, although this is not a matter of great importance; many constructions were built in which the second box lay firmly across the open end of the first. Nevertheless, this method of construction gradually occurs less and less.

Piling higher boxes on lower ones may be done from the ground or from the protruding edges of the *lower* boxes, but also in such a way that the animal, standing on the *topmost* one, pulls the next box up to it. The former proceeding is generally more practical, since the architect does not stand in his own way, as he easily may in the latter case. The animals all adopted it at the beginning. But in the building activities to be described later, in which the whole company joined, too much depended on keeping possession of the top, and so the second proceeding become customary.]

Sometimes it seems advisable to take one of the facts developed by observation, and demonstrate it in sharp outline by an extreme test. For this purpose the animals were confronted with the following situation: the objective is placed very high, a box lies near by, but the ground underneath the objective is covered with a heap of average-sized stones on which a box can hardly be placed firmly. (11.4.1914) Chica gets up on the stone-heap and endeavours in vain to reach the objective with her hand, and later with the stick; she does not pay any attention to the box, and, after a short time, not even to the objective. A second experiment, several hours later on the same day, proceeds in exactly the same way. This tells us nothing. It seems to me completely impossible that Chica should immediately see the stone-heap as an obstacle, as she never reached such clearness of conception in much ruder obstacle-experiments ; in any case she would at least make a try with the box.   The test with the most intelligent of the animals, Sultan, had an entirely clear result, in the same situation and on the same day.   He immediately pulls the box on to the stone-heap, but does not succeed in making it stand up ; he drags a big cage from a distance, tips it onto the stones, sets the first one on top of it, and reaches the objective after fifteen minutes of very hard labour, though on a construction that stands crookedly up in the air.   The stones are now heaped up into a pointed pyramid.   But this time Sultan, by a series of lucky accidents, fixes his box onto the heap in a certain way in a few minutes, and again reaches the objective.   At the third repetition—the pyramid having been built up again—he is not successful, and soon gives up his efforts.   He did not make the least attempt, during the experiments, to move the stones and clear a level foundation.

On the following day the stones are replaced by a number of preserve-tins which are laid undereneath the objective in rolling position.   Sultan immediately seizes the box and attempts to put it on the tins, whereat the box rolls off to the side over and over again.   After fussing about with the box for some time, he pushes the tins (accidentally) a little sideways from the objective, so that a free place is made between them, big enough to place the box perpendicularly.   But he makes further hard efforts to stand the box on the tins without paying the least attention to this free place.   Nothing in his behaviour indicates any endeavour to remove the rolling tins, although he could do it in a few seconds without the least trouble.   Finally the box is put accidentally on the ground and partly on the tins, aslant it is true, yet fairly firm, and Sultan reaches the objective.   [The experiment with Sultan becomes all the more important as this animal takes out of the box the stones which weigh it down the moment he realizes that it cannot be moved ; he thus removes obstacles which he understands to be such. The same earlier experiment also showed that the chimpanzee did not have so much respect for obstacles set up " by the master " that he would not remove them. *This* is an anthropomorphism. Sultan does not consider at all how it comes about that strange objects lie under the objective, and as far as respect goes, he generally reserves that for the moment when, after an offence, the sad consequences actually occur ; unless it is a matter which has been frequently forbidden, such as climbing along the wire-netting of the roof ; which eventually did not happen often in my presence.]

In March, 1916, the same test was accidentally successful, Grande being the animal experimented with. Chica had in vain jumped for the objective with a short stout tree-trunk, and had then left it lying under the goal. Grande began to build, and at first on free ground ; but when, on fussing about with the boxes, one of them tumbled under the objective and thus fell on the trunk, the animal changed its plan, and chose this box as a base. She took all sorts of pains to erect a structure on it, but, all the while, the foundation kept tipping and toppling on the trunk. Grande threw not one glance at the obstacle, any more than Sultan had at the tins.

According to these results, one can construct *a priori* a type of further observations. When the chimpanzee solves problems genuinely, which are only problems of " rough distance " from the goal, and at the same time hardly possesses or learns anything of our naïve statics, " good errors " are bound to occur, in which the animal makes real attempts to conquer the distance better—that is the good in it—but unconsciously aims at a static impossibility—that is the error.

The first of these good mistakes was observed in only two cases ; it has a startling effect. (12.2) Chica tries in vain, in the first experiments, to attain the objective with one box ;

she soon realizes that even her best jumps are of no avail, and gives up that method. But suddenly she seizes the box with both hands, holds it by a great effort as high as her head, and now presses it to the wall of the room, close to which the objective hangs. If the box would " stick " to the wall, the problem would be solved ; for Chica could easily climb up and reach the goal by standing on it. In the same experiment, later on, Grande puts a box under the objective, lifts her foot to climb, but lets it drop again, discouraged, when she looks up. Suddenly she seizes the box and presses it, still looking up towards the objective, to the wall at a certain height, just as Chica had done. The attempt at solution is genuine : the sequence of movements, from " lifting up the foot " to " pressing the box to the wall " contains an abrupt break between " dropping of the foot " and " seizing the box " ; and the proceeding " seizing—resolutely lifting it to about 1 m. high— pressing it to the wall," forms one single whole. Exactly the same applies to Chica's behaviour. It would be a wrong interpretation to say that the animals wanted to knock down the objective with the box. If that were their intention, they would deal with it quite differently, make different movements with it, and would lift the box straight up in the direction of the objective, not press it sideways to the wall, as both did from the very beginning. I will refer again later to this proceeding, for once really containing naïve statics, even though chimpanzee-like and extremely primitive. [One might think that Grande was imitating what she had seen Chica do, but this seems very improbable to anyone more familiar with the chimpanzees' power of imitation. Furthermore, Grande's procedure is that of a genuine attempt at solution, and nothing would be changed if she had copied it ; *it is most difficult for chimpanzees to imitate anything, unless they themselves understand it.*]

If the chimpanzee cannot reach his objective from a

box placed flat, he often turns it upright, after measuring the distance with a glance. There is a further development of this, which only has the defect that it is not compatible with the laws of statics. The animal stands on one box and places another in front of him and on top of the first, but a glance at the goal shows that the distance is too great. Then the upper of the two boxes is turned and turned again out of its position of equilibrium and " diagonally " (cf.

Fig. 10) ; the animal tries meanwhile with grave concentration to ascend the heightened pinnacle. This attempt at solution can be repeated *ad infinitum* as the box certainly moves under the ape's hands, but remains in balance itself, to a certain extent, without much effort on his part. With an amazing stubbornness and minute care, Grande repeated this " good error " for years.

FIG. 10.

To the preceding two a third example may be added, which, though not concerned with building, contains nevertheless a problem of statics. Chica tries to combine her jumping-stick procedure with building ; she either begins her lightning-like climb from the top of the structure while the pole stands beside it, or she tries to prop up the pole *on the boxes*, if the building is firm enough, though, of course, she does not control it visually. If the boxes are so arranged that the top one lies with the opening uppermost, then the highest portions are the narrow edges of the boards composing its sides. *So Chica does not put her pole inside the open box*, but with all care *on its highest portion*, that is, on a point at the edge of the box, a surface about 15 mm. wide. Fortunately the pole always slips down before she has begun to climb properly, or she might easily have a bad fall. She does everything to invite this catastrophe, and always props her pole on the edge. This is a " good " error, arising from understanding of some factors (height and approach to the objective) and complete blank innocence of others (statics).

The setting up of a ladder is, as a task, so like the piling up of boxes that I will pass on to it here. In both cases, when once decided these tools shall be used, there arises the special problem of making them ready for use : a quite independent task of arrangement and statics. However, the ape's treatment of ladders brings out two points not conspicuous in the problem of the boxes.

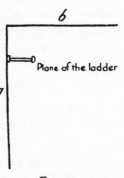

FIG. 11.

When Sultan first made use of the ladder (instead of a box or a table, see p. 48 above) his handling of it looked very strange. Instead of leaning it against the wall near which the fruit was hanging from the roof, he set it up in the open space directly under the objective in a vertical position, and tried to climb up it. If the observer is already acquainted with the animal's habits, he will at once realize that the ladder is here used as a *jumping-stick*. The chimpanzee tries to use this long wooden frame in the same way as sticks and planks. As this method meets with no result, it is altered. Sultan leans the ladder against the neighbouring wall *a* (cf. Fig. 11), but quite differently from the way we do, so that one of the uprights rests against the wall, the plane of the ladder extending into the room. He then ascends the ladder. As the objective is hung from the roof close to one of the corners of the room, and the animal in mounting the ladder has the other wall *b* close before him, he succeeds on the lower rungs in preserving both the ladder's equilibrium and his own, by resting one arm against the wall *b*. But before he reaches the prize, the ladder falls, and after Sultan has had several such tumbles, he lies still for a while in annoyance. Then he returns to the task, and after long trying, finds a position more similar to that in human use in which he succeeds in climbing the ladder and securing the prize. But in this case, as in the previous efforts, he gives the impression of not aiming at placing the ladder against the wall in a human fashion, but, as far as possible, *fitting it to the wall*, while remaining more or less under the objective. The first tendency is the more pronounced and sometimes predominates entirely : therefore, even when the ladder is successfully used, it stands much *too vertically* for our human requirements of statics.

Grande, the exceedingly indifferent acrobat, did not care to do pole-jumping, and so her use of the ladder on the first occasion differed entirely from Sultan's. She had not been present at his test. (Date : February 3rd.) The objective was again suspended from the roof near a corner of the room. Grande brought the ladder to the spot, laid it horizontally against the wall (edgeways up) and tried to reach the objective by jumping from the upright that lay topmost. She had only recently been inititated into the use of boxes ; in the absence of such an article, it is easy to see that she places and uses the ladder as a sort of *defective box*, that has to be propped against the wall. But in the next experiment she lifts up the ladder as Sultan did, in such a way that one upright is propped against the wall and the rungs protrude vertically from it into space. The end of the upright got just enough support against the rough wall to keep it in position, but as Grande began her ascent, it slipped of course and so did she. Nevertheless she often repeated the effort, keeping to the same position of the ladder, until at last and purely by chance, the upright was caught and supported by a little roughness in the wall, long enough for her to climb up (the ladder almost in the air, to our notions) and reach the goal. I repeated this test with Grande three months later (May 14th). She set up the ladder in almost exactly the same position as before, only at a slightly less vertical angle. The caution and dexterity with which she balanced the swaying of the ladder, by movements of her own body, were wonderful and admirable, for as before, the ladder was precariously supported by the end of one upright only, and the whole proceeding looked almost super-static.

Sultan kept to this procedure till 1916. As Chica also prefers this position (and but rarely presses the flat side of the ladder against the wall) this manner of placing it can hardly be mere conincidence. Equally it will be no coincidence that the normal, human method of placing a ladder in position was never perfectly carried out—i.e. unmistakably and at once, as a real solution.

The resulting conclusions may be summarized thus :

1. If we exclude Sultan's attempt to use the ladder as a jumping-stick, we must admit, from the other efforts, that chimpanzees do possess a very modest understanding of statics, and that we can only speak of an *almost* absolute lack of insight in this respect. Both Grande and Chica press a box that is too low, sideways against the wall ; they do not try to suspend it *in vacuo*. Sultan, Grande, and Chica try to put the ladder into contact with the wall, as soon as they perceive the need for firmness, but at first contact is purely *visual* ; and, therefore, not much depends on whether, in their subsequent attempts, an *actual* and practically useful contact is established as well, as long as the ladder will somehow stick to the wall. Even in their procedure with the jumping-stick the same point could be noticed : none of the animals tried to hold too short a stick or a second one, added for length (compare Rana, p. 124 seqq.)—simply higher up *in the air* ; the end must be in touch with something or,

at least, appear to be in optical contact. Thus the dangerous
venture of Chica in putting the end of her pole on the narrow
edge of the open box showed not only lack of clearness in
statics, but also, by the very minute attention and precision
with which she placed the pole just there, and not blindly
in the air, a plain static need.

But the pressing of the boxes against the vertical wall
proved again that this need has not evolved much beyond
visual and physical contact. The placing of the ladder
has certainly been decided by the urge to bring about visual
contiguity between ladder and wall, and therefore is not the
outcome of mere trying around ; but since only this visual
factor comes into consideration, the procedure remains odd
to our eyes. The ladder which is in contact with the wall
by the length of one upright, or by the whole face of the rungs,
is *optically in closer contact* than if it were supported at four
points—the two extremities of each upright—as in our human
fashion. It is then statically rightly placed, but probably
appears to the chimpanzee to be " not firm ", just as *his*
favourite position does to *us*. Unfortunately this visual
factor, too, is never fully evolved : in box-piling there is no
attempt at a real fitting of forms, and even " rough contact "
is ignored to some extent when, for example, boxes project
considerably beyond their pedestals. There is probably, in
problems in which full insight will not in any case be attained,
a tendency to neglect even the possible degree, and so the
ape merely " tries around." In the case of the ladder the
task is a little easier : the relation " homogeneous wall—
simple total shape of the ladder " is more easily comprehensible to an ape than the relation of two boxes. Here,
a certain statics of contact is quite indisputable, though it
varies from human ideas and is very " unpractical." The
particular position of the ladder with one upright against
the wall and the rungs projecting vertically into space is

probably determined by the *position of the fruit*. Had the
first tests in this series been arranged with the objective
attached to the wall, pressing the rungs of the ladder to the
wall would perhaps have been the only procedure adopted.[1]

[As this essay treats as little as possible of theory, I will
give only a brief suggestion of the manner in which the habits
of the chimpanzee positively hinder the evolution of statics.
We know[2] that in human beings the absolute visual orientation in space which makes complete reversal of forms appear
as a strong alteration, develops gradually in children. The
hypothesis that this (normal) *absolute spatial orientation*,[3]
this fixed " above " and " below," is a product of the *habitual*
upright posture of our heads, appears plausible, whether we
wish to attribute the formation of these facts to " experience ",
or (like the author) are inclined to admit a direct physiological
influence of gravity upon the optical processes in certain
parts of the working nervous system (as in this upright posture).
In any case, we should not have developed *this absolute orientation in space* to such an extent, if we, like the chimpanzees
held our heads just as often in other positions as vertically
erect. If we consider the fundamental dependence of our
statics on the generally firm orientation of " above " and
" below ", the " vertical " and " horizontal " (a child too
has no statics as long as it lacks this absolute orientation), it
will be evident that the chimpanzee lives under very unfavourable circumstances to the development of statics.

On the other hand, his natural life is eminently calculated
to exercise the functions of the labyrinth and cerebellum,
and to make him so muscularly dexterous and agile that the
least expert acrobat among chimpanzees need not fear human
rivalry. Thus in the manipulations with ladders and boxes,

---

[1] Subsequent tests about this point gave no definite results.
[2] W. Stern, *Zeitschrift fur angewandte Psychologie*, 1909 ; F. Oetjen,
Zeitschr. f. Psychol., vol. 71, 1915.
[3] M. Weitheimer, *Zeitschr. f. Psychol.*, vol. 61, p. 93 seqq.

he lacks a powerful incentive to the development of statics, for he is physically able to cope with structures to which no human adult would trust himself.]

2. When an inexperienced observer comes into contact with the chimpanzees and wishes to test them in any way, his method is, very often, to give them carefully designed and specialized human implements, e.g. ladders, hammers, tongs, etc., and to inquire whether they utilize these. And then, if such an inexperienced observer sees a chimpanzee using a ladder, he is amazed at the high degree of intelligence and development displayed. But we must quite distinctly understand that the chimpanzee is *not using a " ladder " in the human sense* of the term (which connotes both a special *orm* and a special *function*), and that, for the ape, a ladder has no particular advantages over a strong plank, a pole, or a tree-trunk, all of which he utilizes in much the same way,[1] for he only apprehends the rough qualities of the whole object and its most primitive functions.

But the observer is far less impressed by the utilization of tree-trunk, pole, or plank, just because he was dazzled and misled by the *external* " humanness " of the chimpanzee's employment, of a " real ladder ", though the trunk, pole, and plank are, for the chimpanzee, absolutely equivalent to the ladder. We must be very careful in this case, as always in investigating the ape's nature, *to distinguish between the external impression of humanness—possibly only due to the instrument used—and the degree of insight and the level of achievement displayed*. The two are not, by any means, necessarily parallel. I must explicitly state, in order to dispel any misconceptions, that I do not recognize any differ-

---

[1] This, although he certainly *sees them as " different things "* and, as this whole book proves, does not simply pass through diffuse streams of phenomena. (Volkelt, *Vorstellungen der Tiere* (1914)—where lower forms of animal life are considered). I admit that the objects perceived by the chimpanzee have not all the qualities of *our* objects.

ence in value between the employment of a ladder and of a jumping-stick by the chimpanzees ; and I consider that there is only a minute difference in that respect between the placing of a ladder *under the objective*, and the same procedure with a strong plank. Ladder and board are both utilized in the same manner and are *practically the same* to the chimpanzee, as he grips with his feet. For us humans, they are quite different, and while the chimpanzee's jumping-stick would be a wretched implement for most human beings, it is even more convenient for the chimpanzee than the ladder. External resemblance to human procedure is no criterion here.

We must always first consider the *function* of the tool, the purpose and manner in which it is used *by the chimpanzee* ; we must analyse and determine what qualities and properties he *realizes*. And, having learnt the *range of functions*, within whose limits the chimpanzee is able to understand the utility of any object, we shall prefer to investigate his achievements and methods of arriving at his solutions in this clear and simple domain, instead of bringing him into contact with the complex products of human craftsmanship. For in such products—e.g. even in ladders, hammers, tongs—there are combined a great number of delicate functional points of view. The ape will always leave uncomprehended and unrecognized full half of what, to us, are the essential requisites in such a tool. He will make, on the one hand, an impression of dullness or confusion, because he uses the tool wrongly, and, on the other, he will look imposingly " human ", just because he " handles ladders, hammers, and tongs." The experimental tests furnish clearer and more valuable results, both for the estimate of the chimpanzee's stage of development and for the psychological theories one wishes to base on such research, if we do not employ as " material " the complicated tools of human invention, but confine ourselves to the most primitive objects—primitive both as regards form and function.

Otherwise we only confuse both the animals and ourselves as observers. Only as long as the region of simple intelligent treatment of the surroundings is not even superficially investigated, can one fail to see that we must study the simplest functions which can be grasped with insight, before the animals are overwhelmed with whole conglomerations of problems at once.

The position is somewhat different in respect of another class of problems: when there is no longer a question of discovering what the chimpanzee is able to achieve by his unaided efforts. When once we are to some extent in a position to judge this, we can pass on to further tests with more intricate conditions and material, in which we offer him all possible teaching and assistance, in order to see how far he learns to understand. We human beings, too, did not discover all the methods of acting intelligently in a day, but learnt much by the aid of instruction. And so it would be a significant problem to solve, whether the chimpanzee could learn to comprehend the human use of the ladder, or whether eventually he could —with human help—realize the essential function of a pair of tongs.

Supplement: Building in Common.

After the chimpanzees available in our researches were already familiar with the process of piling one box on another, the whole group was often afforded opportunity to build up boxes towards an objective suspended at a considerable height in the playground. In time this became a favourite amusement. But we must not suppose that this " co-operative building " represents any systematic collaboration, with any strict division of labour among individuals. This is, rather, the procedure : The objective is hung in position, and the assembled chimpanzees gaze around for material to use as tools. In a minute they have all rushed under it, one with a pole, another armed with a box ; sometimes they drag their tool along the ground, but Chica prefers to lift her box up in her arms or to balance her plank on her shoulder like a workman. Then several of the animals want to ascend at the same time, each behaves as if he alone were about to " build ", or had himself erected any " pediment " that may exist, and wished to complete the structure quite unaided. If one ape has already begun this constructional exercise, with others building close beside him, as frequently happens, a box is unhesitatingly pilfered from the neighbour's store and the rival architects come to blows ; this is apt to interrupt the progress of the work, as the higher the structure, the keener the competition to mount it. The result is generally that the object of the struggle is itself destroyed in the struggle —knocked over in the mêlée. So the apes have to start again from the beginning, and thus Chica, Rana, and Sultan often give up the labour and struggle, while Grande, the oldest, strongest, and most patient of the four, is left to complete it. In this way she has gradually acquired the most skill in building, although the more impatient animals, Chica and Sultan, are distinctly superior to her in intelligence.

It is only rarely that one animal *helps* another, and when this happens, we must carefully consider the meaning of such action. As Sultan was much more expert than the others, in the beginning, he was often obliged to be present without helping, as I wished to ascertain the capabilities of the others. In one of the illustrations (Plate IV) we may observe his attentive interest. (He is below, on the right-hand side of the picture.) If the observer's vigilance is at all relaxed, and the veto on building not continuously renewed, Sultan does not venture to enter fully into the work, but he cannot keep from " lending a hand " here and there, supporting a box that threatens to fall under some adventurous and decisive effort of another animal, or otherwise taking a less important part in the work. (Compare Plate VII, which is a reproduction from a cinematograph film : Sultan is steadying the box, which moves under Grande's weight.)   On one occasion when we had forbidden him to participate in the building, he could not keep to the rôle of passive spectator, when Grande had piled one box on the other, and was still unable to reach the prize.   He quickly fetched a third box from a distance of about twelve metres and put it close to the pile ; *then he squatted down again and watched*, although he had not been reminded of my prohibition by either word or gesture[1]. But we must guard against misconceptions : Sultan's motive is not the wish to help his fellow, at least not predominantly. When we watch him, squatting beside the other animal, following all Grande's movements with his eyes and often with slight sketchy movements of arm and hand, there can be no doubt that these proceedings *in themselves interest him*, and to a very high degree ; that he follows and " feels " the movements himself, and all the more keenly as they grow more difficult and crucial.   The " help " he offers at the critical moment is simply *a heightening of his already indicated participation in the process* ; and interest in the other animal can play only a very secondary part, for Sultan is a pro-nounced egoist.   In the second part of this work I hope to show the extent of this " participation ", and the compelling urge that seems to overwhelm the chimpanzee who is a spec-tator.   (Cf. on Plate VII, in the foreground : the animated gestures of Konsul at the critical moment : on the cinema film this is, of course, much more graphic.)   We are all acquainted with similar states of mind.   It is difficult for anyone who, as a result of long practice understands any form of work, to stand aside while another bungles it : his fingers itch to intervene and " do the job ".   And we human

---

[1] I have already said what is necessary on the subject of " reading into " and " anthropomorphism ".   There was no ambiguity whatever in Sultan's behaviour on this occasion.

GRANDE
PLATE VII. KONSUL   SULTAN   CHICA
      BUILDING ALTOGETHER (BUT NOT IN COMMON)

beings, too, are far from wishing to help such a bungler from motives of pure altruism (our feelings towards him at the moment are not particularly cordial). Neither do we seek some external advantage for ourselves : *the work* attracts and dominates us. Sometimes I think that in these traits of character the chimpanzee resembles us even more closely than in the realm of intelligence in the narrower sense ;— but we should be on our guard against a mere *intellectualist* interpretation of them. (A fine example of this resemblance in minor traits is the habit of " passing on " a punishment that has been suffered to an habitually unpopular or un-congenial animal : Sultan often does this to Chica.)

Sometimes the behaviour of the animals strongly resembles collaboration in the strictly human sense, without, however, entirely carrying conviction. (Date : 15th February.) The little ones had made repeated efforts to reach an elevated objective, without success. At some distance stood a heavy cage, which had never before been used in the tests. Suddenly Grande's attention was caught by this cage ; she shook it to and fro, to turn it over and roll it towards the objective, but could not move it. Rana forthwith came up, and laid hold of the cage in the most adequate way, and the two were in the act of lifting and rolling it, when Sultan joined them and, seizing one side of the cage, " helped " with great energy. Alone, none of the three could have stirred the cage from its place, but under their united efforts—which were " timed " together perfectly—it rapidly approached the goal. It was still at a little distance when Sultan bounded upon it and then, with a second spring, secured and tore down the fruit. The others received no reward, but then, they had worked for themselves and not for Sultan, who had good reason to take a sudden dash forward, for otherwise he might have been " done out of it.". Rana certainly understood Grande's intentions when she first began to move the distant cage, and

M

took a hand in her own interests just as Sultan did. As all three had the same aim, and as the moving cage prescribed to all of them the form of procedure, the box was rapidly rolled on its way.

The following examples are pendants to Sultan's behaviour when he saw others building and was excluded from the competition. As he is, generally speaking, much in advance of the other animals, he is sometimes permitted to be present when they are undergoing tests with which he is already familiar. He pays close attention—as in the building experiments—but is not allowed to take part. If the test is one in which the animal under observation is on the opposite side of a grating—for Sultan watches from *outside* the bars—and the objective is outside on the ground, no stick being available, Sultan watches the other animal's ill-adapted efforts quietly for a time. Then he disappears, to return with a stick in his hand. With this stick he scrapes sand together, at a distance from the objective, but near the bars, or pokes through the bars. If the other ape tries to grasp it, he pulls it away seemingly to tease, and thus there develops a to and fro game, *which tends to leave the stick in the neophyte's hand, if no prohibition on my part intervenes.*

In one of these tests the neophyte could have supplied himself with a stick by breaking up the lid of a box which stood near the bars. Sultan was sitting outside, but the other chimpanzee failed to solve the problem. Suddenly Sultan began to shuffle towards the bars, until he was quite close to them. He cast a few cautious glances at the observer, stretched his hand between the bars, and tore a loose board from the lid. The further course of this test was an exact repetition of the one just described.

In both cases, as in the building examples, Sultan's behaviour shows no trace of " altruism ", but, though he takes no part in the procedure, we feel his complete comprehension of it, and his imperative impulse to *do something* towards the solution which remains so long undiscovered.

It is clearly proved by the following instance that he really sees the task to be carried out, *from the standpoint of the other animal.* I was endeavouring to teach Chica the use of the double stick. I stood outside the bars, Sultan squatted at my side, and gazed seriously, slowly scratching his head meanwhile. As Chica absolutely failed to realize what was required, I finally gave the two sticks to Sultan, in the hope that he would make things clear. He took the sticks, fitted one into the other, and did not himself appropriate the fruit, but pushed it, in a leisurely manner, towards Chica at the bars. (Had he been very hungry, his behaviour would probably have been quite different.)

Mutual obstruction is more frequent than co-operation. Tercera and Konsul do not take part in the building operations ; they sit on some point of vantage, and watch the others at work. But when the building is in full swing, they give striking proof of their comprehension. They love to creep up behind the back of the busy architect, especially when he is perched precariously high, and, with one vigorous push, knock both building and constructor to the ground. They then flee at top speed. Konsul was a master of this game as well as of every grotesque contortion. With an expression of comic rage, stamping, rolling his eyes and gesticulating, he prepared his fell design behind the innocent constructor. It is impossible to describe happenings of this sort ; I have seen observers shed tears of helpless mirth as they watched them.

The emotional foundation of this behaviour is a little difficult to understand ; in that described below it seems clearer and also has been frequently observed. One of the animals has just completed his building; then suddenly another, the redoubtable Grande for instance, approaches, with unmistakable intent to use the first animal's efforts for her own advantage. A pitched battle seems inadvisable, but the smaller animal does not at once take to flight, leaving the field clear. Instead, he sits on the edge of the topmost box and slides off it in such a way that the whole structure overbalances and collapses. This proceeding differs totally from that usually adopted in descending and must be intentional ; flight follows and rage on the part of the outwitted aggressor.[1]

[1] A. Sokolowsky made observations on a number of anthropoids in Hagenbeck's Zoological Park. I find that other investigators cast doubts on some of the achievements of the apes as recorded by him. It is certainly true that an expert psychologist would choose his expressions in the description of the subject more carefully, and show more reserve in his comments. But, after my investigations so far, I find the bulk of the facts recorded quite probable, and he correctly recognizes that under certain circumstances anthropoids act with insight. And let us not forget that Sokolowsky was the first person to suggest the psychological study of anthropoids on an experimental basis, in special institutions on account of the deficiencies of all occasional observation.

VI

DETOURS WITH INTERMEDIATE OBJECTIVES

In some of the instances described above, there has been quite a long detour to make. Sultan spends considerable time in gnawing the end of a board that he wishes to fit into a hollow tube ; and yet this gnawing of wood at the end of a stick is an activity which, considered separately, is meaningless in relation to his objective. It is not by any means easy for the observer to regard it in this way. What we see is rather, " Gnaw, gnaw, see if it fits into the tube, gnaw, fit, etc.", an objectively-connected sequence. What becomes of the experiment, if we carry this principle a step further ? In the case of simple preparation of a tool, and so in our example, the part " preparing " (gnawing) is still somewhat closely bound up with the further procedure (insertion and employment) by the fact that the auxiliary action has a direct bearing on the tool-material. If we now try to find tests with parts of greater independence, we arrive at experiments in which the animal has to draw into the situation a preliminary " minor " objective, before he can reach his final objective. This auxiliary objective must itself be approached indirectly, if the final objective is to be attained. On the other hand, if the preliminary process, up to the attainment of the minor goal, be considered quite by itself, and apart from what follows, this first detour shows less connection with the final goal and externally can be distinguished as a separate procedure. Experience shows that we get a particularly strong impression of intelligent behaviour when detours are made as *one* action, which in their separate phases lead so far away from the goal, but which, considered as wholes, correspond exactly to the situation.

(March 26th) : Sultan is squatting at the bars, but cannot reach the fruit, which lies outside, by means of his only available short stick.   A longer stick is deposited outside the bars, about two metres on one side of the objective, and parallel with the grating.   It can not be grasped with the hand, but it can be pulled within reach by means of the small

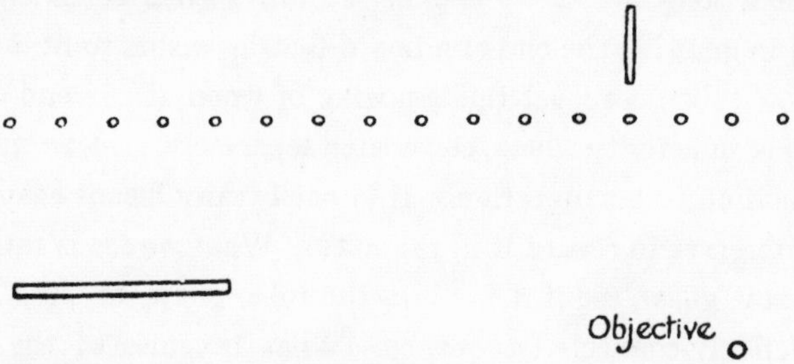

Objective

FIG. 12.

stick (see Fig. 12).   Sultan tries to reach the fruit with the smaller of the two sticks.   Not succeeding, he tears at a piece of wire that projects from the netting of his cage, but that, too, is in vain.   Then he gazes about him ; (there are always in the course of these tests some long pauses, during which the animals scrutinize the whole visible area).   He suddenly picks up the little stick once more, goes up to the bars directly opposite to the long stick, scratches it towards him with the " auxiliary ", seizes it, and goes with it to the point opposite the objective, which he secures.   From the moment that his eyes fall upon the long stick, his procedure forms one con- secutive whole, without hiatus, and, although the angling of the bigger stick by means of the smaller is an action that *could* be complete and distinct in itself, yet observation shows that it follows, quite suddenly, on an interval of hesitation and doubt—staring about—which undoubtedly has a relation to the *final* objective, and is immediately merged in the final action of the attainment of this end goal.

(April 12th) : Nueva was tested in the same manner. The little stick was placed on her side of the bars, exactly opposite to the objective, and the longer stick outside the bars, somewhat nearer than the objective, and about one and a half metres to one side. As Nueva was already at that time seriously indisposed and had very little appetite, she soon gave up her efforts, when she found that she could not angle the fruit with the short stick. Then some specially fine fruit was added to the prize, and she approached the bars once more and gazed around. Her eyes fixed themselves on the larger stick ; she took the little one, drew the larger within reach and secured the fruit. The whole action could not be more distinct and coherent.

Grande was tested at a much later date (March 19th, 1916). She scraped the small stick vainly in the direction of the fruit, then took no notice of the experiment for a time, only to return to it and repeat her efforts with the little stick. Then she sat motionless for a minute by the bars, opposite to the fruit. When her eye fell on the larger stick at the side, she stared at it, but still remained motionless. Suddenly she jumped up, pulled it in with the little stick, and then fetched the fruit in with it. The solution of this experiment is not at all " obvious " : we repeated it a few minutes later, but this time we placed the larger stick in the spot where the objective had been before, only nearer, and the objective in the same place as the big stick in the first experiment, but somewhat further away. Grande again made fruitless efforts with the small stick, and then became indifferent. We called her ; she came up to the bars ; squatted opposite the fruit, and looked about until her gaze was arrested by the longer stick, when she repeated her former correct procedure. It is always a sign of extreme difficulty if a rapid repetition of the test does not result in a rapid repetition of the solution just found.

When Chica was tested, a mistake was made in arranging the experiment, for while she was making vain efforts with the smaller stick, she saw the other, snatched at it, and actually succeeded in reaching it and with it the fruit. The second time we put the long stick further off ; she secured it with the smaller one, and so on.

The less gifted animals of our group showed the value of this achievement more definitely. (April 1st, 1914) : Tschego worked hard with blanket, straws, and handfuls of straw to secure the prize, as well as with the little " auxiliary ", but, of course, without result. The large stick, which lay a trifle to one side, on free ground, plainly visible and easily accessible by means of the other, was not heeded for a moment, and, after waiting more than an hour, we had to break off this experiment. The same negative result followed a repetition in 1916.

Whereas one might think that Tschego had not " noticed " the long stick, Rana in her experiment had not even this excuse (19.3.16). Clumsily[1], she angled with the auxiliary, and then approached the larger and *stretched her hand out towards it.* Her whole behaviour could thus be expressed in human speech : " I shall not reach the objective with the little stick—outside there is a long stick which my hand cannot reach." She did not for an instant seem to conceive the auxiliary stick (which remained inseparable from the objective) as an instrument for securing the longer stick. Finally I gave her some assistance. In order that she might with greater ease connect the little stick with the long one, I pushed it away from the objective, and nearer to the big stick, while Rana was looking in another direction. I continued this till the small stick was quite close to the big one. *Nevertheless, as soon as Rana had seized the short stick, she hastened back*

[1] There seems to be in apes a high positive correlation between intelligence and dexterity.

to the point just opposite the fruit, and angled for it with the totally inadequate bit of wood. The detour " short stick—long stick—fruit "—simply does not arise with this animal. She reminded me of the hens, that persistently charged straight at the wire-netting in front of a coveted morsel, although a very short " way round " would have brought them to it at once ; just in the same way, Rana scrapes and stretches vainly towards the fruit which she could have secured with so much less exertion. It almost seemed as though the short stick were attracted by an unseen but strong force into the primary critical distance " goal—bars ", and therefore did not come into consideration at all for the secondary distance " long stick—bars."

In the next test the fruit was again placed outside the bars ; inside a stick was hung from the roof some distance from the bars and a box placed at one side. The fruit could be reached with the stick, but the stick only through the aid of the box. (4.4.14) Sultan began his treatment of this problem by a foolish mistake, and dragged the box to the bars, just opposite the prize. After moving the box to and fro a while, he left it alone, and began in a more careful way to look about him (obviously seeking an implement), and now saw the stick hung from the roof. At once he made for the box, pulled it under the stick, stepped up, tore down the stick, hurried to the bars, and pulled in the fruit. From the moment he caught sight of the stick, his actions were perfectly definite, clear and continuous. The time that elapsed was at most half a minute—including the angling for the fruit with the stick.

Chica did not arrive at this solution though the stick was hung in position in her presence (April 23rd) and also touched and moved in order to attract her notice. (May 2nd) : The stick was again fastened to the roof, while Chica looked on. She took no notice of it, but tried to reach the fruit with a flaccid plant-stalk and then to tear a plank loose from the lid of the box ; finally she tried to reach the fruit with a straw. Then she appeared to lose interest and began to play with her companion, Tercera ; the stick seemed no longer to exist for her. Suddenly some one in the distance called loudly ; Chica started and looked round, and, in so doing, caught sight of the stick. Without any hesitation, she went towards it, leapt straight upwards twice, and unfortunately for the purposes of investigation, succeeded in reaching it, as a slight rise in the ground assisted her efforts.

It is to be remarked that, in this case as in Sultan's, the stick was noticed and secured *as an implement*, although some time had elapsed since the objective had occupied her attention. The interval was about ten minutes in the case of Chica, who played with Tercera during this time. Yet, as soon as she perceived the stick, she secured it, her behaviour being not at all influenced by any glance towards the objective.

Immediately after this first test, the stick was hung to another part of the roof, where it was not possible to reach it from the ground. The box remained in the centre of the room. Chica sprang at the stick with indefatigable persistence, but in vain. There was obviously no connection between box and stick for her, for she squatted on the former more than once, to get her breath again, and did not make the least attempt to place it under the stick. But presently the reason became evident ; as soon as Tercera, who had been reposing on the box all this time, descended for some purpose, Chica seized it and pulled it forward, mounted it and seized the stick. But when she had it in her hands at last, the major objective was evidently forgotten for a moment, for she stood for a few seconds, turning her back to the bars and the fruit, and gazed at the stick helplessly. But her orientation was renewed without stimulation from without ; she suddenly turned round and hastened towards both bars and objective

—clearly so that the turning was already part of the movement towards the objective.

One can hardly assume that Chica had already seen the box as an implement, to be used for the mid-objective (the stick), while Tercera was still lying on it. To judge by the conduct of the animals at other times, she would in that case have tried to remove her friend, by pleading and complaining and by pulling at her hands and feet, and would at least have made an attempt to move the box, in spite of its weight. (Compare also p. 60 above, where a similar predicament arose and the same animals were involved.) It is only when the box is freed of Tercera that it is recognized as an implement; as the seat on which she is sitting is not thus recognized. The experiment also shows that obstinate attempts to gain the mid-objective, although prompted by the desire to attain the final one, can displace this to a certain extent, so that at the end of the minor activity there is a check. On the other hand, the stage of development of the chimpanzee can hardly be better gauged than by the way that Chica finds her way back to the complete problem. She has concentrated for a considerable time on the mid-objective, not throwing even a glance at the major one in the meantime, and then after a few seconds of helplessness, gets back her orientation as if with a start. She remains with her back to the major-objective, so that *nothing external* can cause that start, except perhaps the stick in her hand; and yet the mere sight of the stick is, of course, not enough to account for it. (Compare here Sultan's behaviour in the test with the invisible box, p. 52.)

Much more extraordinary did the relation between major- and mid-objective turn out in the same test, with Koko as the subject of experiment. (31.7.1914) As before, he is confined, by collar and rope, within a circle of about four metres radius; the objective lies on the ground out of reach, and the stick also is placed out of reach above, hanging from

a smooth wall, and the box stands at the side (cf. Fig. 13).
Without doubt the experiment starts with a clear solution
in sight ; for Koko grabs hold of the box at once, and pulls
it straight towards the stick on the wall, the use of which, of
course, he knows exactly by now.   But unfortunately he has
to páss the objective on his way, and, as he reaches the spot
where this lies in plain view on the ground, he suddenly turns
at a sharp angle away from his straight and unmistakable

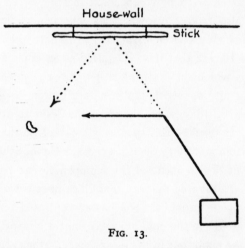

FIG. 13.

path towards the objective, and uses the box as a stick—by
letting its upper end fall on the fruit—and then he pulls ; this
method succeeds.   Exactly the same thing happened when
the experiment was repeated.   Again the box is taken in a
straight line from the starting-point towards the stick, and the
meaning of his action is indicated plainly enough by his
frequent glances towards the mid-objective ; but, on passing
over into the region where lies the final objective, the animal
seems to be turned away sideways toward the final objective,
and from this moment the stick on the wall is noticed no more.

The next day the behaviour just described developed
to a degree of stupidity familiar enough to us from our ex-

perience with Sultan. The start was right ; but the box, once again dragged towards the stick, could not be got past the final objective. On the contrary, Koko was again diverted from the straight path towards the mid-objective, but instead of using the box as a stick, which would have been fairly sensible, he now pushes it with all his might towards the objective, mounts it, and tries to reach the objective from the top, as if the objective were placed high up ; actually, in this situation, the box is only an obstruction, increasing the distance between the animal and the objective. But this utter stupidity of sitting up on the box and trying to grasp the objective is persisted in for only a few seconds, and the attention is turned anew to the stick on the wall. Yet the box, having once been put aside, is left there, as if consecrated to the final objective ; at any rate Koko takes the greatest trouble to reach the stick without using it.

Comparing this behaviour with that at the beginning of the experiment, the difference is so unexpected that it is necessary to find out now whether the use of the box has not, as once before, been suddenly lost sight of. The stick is taken away and the final objective hung in its place ; Koko reaches for it for a moment, then quickly fetches the box, gets up, is unable to reach the goal, jumps down, corrects the position of the box quickly and neatly, gets up once again, and reaches the objective. According to this, the box simply could not be dissociated from the main or final objective, just as before it could not be got past it, although that time it was already on the path towards the secondary objective. For we must completely exclude the explanation that Koko only understood how to use the box in relation to the fruit, and not to other objectives ; even the beginning of the experiment, when the situation had its effect at once in the sense of " put box under stick ", shows clearly that the difficulty is not of so external a kind. Rather it must be a question of " relative value ", of final and secondary objectives in relation to one another, in the sense that the " stronger " main object can draw away to itself the " auxiliary " action bent on the secondary " weaker " object ; while the correct but very indirect path past the secondary to the main objective can arise (beginning of experiment), but is interfered with in that second (as in a short-circuit) in which (at the passing of the region in which the main objective lies) the main objective is dangerously near.[1] In the experiment with Rana described above, the short handy stick is so completely " bound " to the final objective that it is not free for the secondary objective even for a moment, and the parallel here is that the box once connected with the final objective is abandoned as if pre-empted, while Koko cranes himself towards the stick on the wall in vain. (Rana too reached for the long stick only with her free hand).

(6.8) Again the experiment, at first, runs along similar lines. The box is brought as near as possible to the final objective and used as a stick ; for some moments Koko mounts it, though this is nonsensical, but only after the animal has, through vain attempts, got into a rage. His excitement gradually increases, he starts hitting and pushing the box with all his might, then lets it go again and turns to the stick on the wall. After he has several times reached up towards it in vain—the box is wedded to the final goal or the critical distance, as before—he at last gives up the job altogether. As the experiment looks hopeless, it is interrupted for a few minutes, and Koko is left alone. When the observer returns, the box is standing under the nail which held the stick to the wall, the stick is on the floor where the objective was before, and this latter is just disappearing into Koko's mouth. The original situation is at once reset, and Koko

---

[1] I have drawn attention in the Introduction to the theoretical importance of mistakes.

solves the problem without hesitating for a moment, or going off on the wrong path, as before. This was the second experiment of which, after long waiting, I did not see the solution. That this particular test should succeed sooner or later was, of course, to be expected, from his right start at the beginning, and as Koko was completely isolated during my absence—i.e. about three minutes—he must have hit on his solution without outside help. More than this, the repetition shows that it was now completely grasped.

[The day before, a sort of reverse of the experiment had immediately succeeded. The objective hangs from the wall, a stick lies near, a box is placed out of reach. As Koko's arms are very weak, he does not succeed in beating down the objective with the stick ; so, after a while, he attacks the box with the stick, pokes carefully with the point of the stick inside the box (which is open on top), tips it over towards himself so that he can reach it with the tips of his fingers pulls it to him, brings it under the objective, and so on.]

If one tries to make the roundabout way still more indirect and difficult, the tendency to leave it for more direct paths or attempts becomes, of course, yet stronger.

(11.5) The arrangement remains the same, but the box is filled with stones. Sultan looks around for a moment notices the stick on the roof, gazes at it concentratedly, goes to the box, and pulls it with all his strength towards the stick. As he can barely move from the spot, he bends down and, takes out one stone, which he carries under the stick ; he places the block upright against the wall, but after one look up, does not climb it after all. (Here the stone, which should only have been taken out of the third-class objective (the box), was attracted towards the next objective ; in this case the short cut does not hurt.) Thereupon he at once drags the same stone to the bars opposite the end-objective, and tries to push it between them ; obviously the stone is to be used as a substitute for a stick.   But, though otherwise practicable in shape and length, it will not go through the bars.   The rest is clear and simple : Sultan again turns towards the box, takes another stone out, with difficulty pulls the box (still weighed down with two blocks) under the stick, stands it upright (whereupon the last stones accidentally fall out), takes the stick down, comes to the bars with it, and immediately reaches the objective.   The same would have happened from the beginning, if the shorter, but less practicable, paths of action had not arisen so easily and, at least for a time, blotted out other better, but too indirect, round-about ways.

On the whole, we get the impression that there is no progress to be hoped for from these experiments.   Whereas in the examples cited we were able to understand what is happening to the animals, further complications of this kind would probably lead to one " deflection " after another, and so, in the end, make it merely difficult to distinguish the behaviour observed from simple trial-and-error conduct.   It is only when one has carried out many intelligence tests on the apes that one learns to avoid with real dread this vague border-land.

VII

"CHANCE" AND "IMITATION"

THE experiments described heretofore proceed in general very simply and somewhat in the same manner. As those in the next chapter are of a rather different nature, it seems to the point to mention beforehand certain considerations which will ward off some facile objections against the real meaning and value of the facts. This would not be necessary in the case of results of a highly-developed experimental science like physics, in which the meaning of groups of observations cannot long remain altogether a matter of controversy. A system of knowledge which cannot be destroyed stands there firm and clear ; and new discoveries *must connect up* with it in one way or another. No one can deny that we are far from such a happy state of affairs in the higher psychology. Instead of sure and fruitful knowledge, we have, so far, developed for the most part nothing but theories of a very general form, but so indefinite that even their supporters would not find it easy to apply them strictly and in detail to any specific case. The more energetic does the claim become that any one or other of these opinions contains the principle that will explain a great number and variety of phenomena ; and their loose connexion with actual experience, together with the indefiniteness of the assertions, make it the more difficult to decide a conflict by observation and experiments, which is still almost in the region of a battle of faith. At the same time, it is inevitable that such actual observations shall lose in value. They are all too peculiar, too individual to attract the attention already give to any general principles. And the indefinite nature of these principles on the one hand, added to the difficulty of really reliable observation on the other, make it possible for nearly anyone to explain anything. Thus, if, at the outset, there is greater interest in general principles than in facts, the facts must finally seem valueless ; they can be explained as one wills.

In this book, no theory of intelligent behaviour is to be developed. Since, however, we have to decide whether chimpanzees ever behave with insight, we must at least discuss certain interpretations which cannot be accepted without the observations at the same time losing all their value in regard to this question. This will at least prevent any quite arbitrary treatment of the facts, and the direct meaning of the experiments will appear with more force and certainty. Perhaps finally it will be possible to make this meaning rest on its own merits, instead of allowing it to disappear in the solvent of general and indefinite principles.

An interpretation mentioned above runs : The animal solves its task in the general form of " roundabout behaviour," and since it is not born with a ready reaction for each case, it must *develop* a new complex attitude for these cases. The only possible origin for complex action is a great many fractions or parts of the whole achievement, which *separately* are quite natural to the animal; such " natural " impulses occur in great variety, and a certain number of them, which in the play of chance, may happen to follow each other, form, when put in a series, the whole actual course of the experiment. As the actual success or the corresponding pleasant feeling has the effect, in a manner not yet known, of making the preceding movements reproducible in later cases of a similar nature, with the origin of such achievements, their repeatability is also explicable. Like most of these general theories, this one has a certain value for some of the cases arising in animal psychology. Where experience might teach one to doubt, two supplementary principals are usually called in to help. According to the first, the application of a theory up till now so well confirmed must be preferred to the recognition of conflicting facts and the development of corresponding new ideas, and this for the sake of scientific economy. According to the second principle, such behaviour directly arising out of the situation as a whole would be a miracle, which must be excluded *a limine* as contradictory to the foundations of scientific knowledge.

Further discussion of these auxiliary principles must be omitted here. The second declares the insolubility of a scientific problem, whose solution no one has yet properly attempted : why this lack of courage ? The first expresses a misconception (at present widely current) of a correct epistemological proposition, according to which a scientific system, which is near to completion (i.e. half-ideal), tends to acquire the most precise form and strictest simplicity. Neither principle has the right of control over experience, and, where there arises a conflict between these principles and observations, the *former* must give way, not the latter. Evidently the epistemological proposition neither states nor implies that a science which has existed for only a few decades should at any price get along with the minimum of points of view that it has discovered in this early youth ; and one does not get nearer to the ideal state any faster by trying to force the shortest cut to the goal of strictest unity, by proclaiming meagre beginnings as final principles, and by economizing on facts what one does not wish to spend on theories.[1]

The chief thing to be done now is to present the theory of which we have been speaking in such a way that its relation to the intelligence tests described will be easily seen. Let

[1] Cf. *Nachweis einfacher Strukturfunktionen u.s.w.*, *Abh. d. Preuss. Ak. d. Wiss.* 1918, No. 2, p. 40, seqq.

those parts of the " solution " which the animal in theory achieves " naturally " and by chance be : a, b, c, d, e ; generally besides these, and amongst them (without them, too), any others : F, Y, K, R. D, etc., may appear in any order.

The first question is whether a is accomplished with any reference to the fact that b, c, d or e should follow so that they all together produce a curve of behaviour which is adequate to the objective structure of the situation.   Not at all ; for when a occurs, it has as little to do with b, c, d, and e as with F, Y. K, etc., which might follow on a, just as well, and generally will, in whatever permutation ; the succession is, after all, as accidental as the winning numbers in roulette.   What is true of a is equally true of all the other " natural " fractions ; to use a phrase which is beyond a simple analogy and which brings the whole question into connexion with the second principle of thermo-dynamics—they are all *completely incoherent*, representing in a somewhat magnified form a case of " molecular disorder."   If the least thing is altered about this, the whole meaning of the theory is destroyed.

The second question is whether, after the animal has developed the sequence a, b, c, d, e, it *then* begins with a, goes on with b after a, and so on, because these fractions in this order correspond, as a whole, to the objective structure of the situation ? Doubtlessly not.   It proceeds from a to b, and so on, only because the after effects of its former life force it to make b follow a, c follow b, and so on.

Thus the only way in which, according to this theory, the objective conditions of the situation, its structure, can have any effect in the development of the new behaviour, is by a completely external meeting of the objective circumstances and the chance movements of the animal's body ; the situation works, roughly speaking, like a sieve, which lets through only part of the things which are thrown into it.   Apart from this result of the objective situation, which is not of great interest for our investigation, we find that *nothing in the behaviour of the animal arises directly out of the relation of the several constituent parts of the situation to each other ; the structure of the situation in itself has no power whatever directly to determine conduct appropriate to it.*

[According to the theory one has to recognize *one* limitation of the natural reactions caused directly by the situation ; a tendency to keep approximately in the direction of the objective[1]. Round this fundamental motive, generally characterized as a manifestation of instinct, play the actual separate movements, whose spatial range thereby becomes narrower but which in this narrower range remain accidental as before. Since taking the straight direction towards the objective does not achieve much, in the experiments here performed, and is often even inadequate to the situation, I need not dwell longer upon this point, while there is no question of the formulation of a positive theory. The tendency is important enough, however, because it indicates something quite foreign to the principle of chance, but hidden under the harmless phrase "instinctive impulse". (Cf. remarks on this primary direction of action, p. 88 seqq.)]

Now I mention from the very beginning, how in the case of a roundabout-way experiment a practical "result" consisting of single and separate fractions, put together by chance, is sharply distinguishable from "genuine solutions". In these, the smooth, continuous course, sharply divided by an abrupt break from the preceding behaviour, is usually extremely characteristic. At the same time this process as a whole corresponds to the structure of the situation, to the relation of its parts to one another. Thus, for example : the objective is visible behind an obstacle of a certain form, on free ground —and suddenly the smooth and unchecked movement occurs along the corresponding curve of solution. We are forced to

---

[1] In the case of animals directed by their sense of smell : in the direction of the strongest intensification of the smell.

the impression that this curve appears as an adequate whole
from the beginning, the product of a complete survey of the
whole situation. (Chimpanzees, whose behaviour is incomparably more expressive than that of hens, show by their careful looking around that they really begin with something very
like an inventory of the situation. And this survey then
gives rise to the behaviour required for the solution.)

We can, in our own experience, distinguish sharply between
the kind of behaviour which from the very beginning arises out
of a consideration of the structure of a situation, and one that
does not. Only in the former case do we speak of insight, and
only that behaviour of animals definitely appears to us intelligent which takes account from the beginning of the lay of
the land, and proceeds to deal with it in a single, continuous,
and definite course. Hence follows this criterion of insight :
*the appearance of a complete solution with reference to the whole
lay-out of the field.* The contrast to the above theory (parts
put together by chance) is absolute : if there the " natural
fractions " were neither coherent with the structure of the
situation, nor among themselves, then here a coherence[1] of the
" curve of solution " in itself, and with the optical situation, is
absolutely required.

[To anyone who is inclined to regard the above explanations
as detailed trivialities, I would suggest a glance through the
psychological literature of man and animal. These trivialities
should be thoroughly emphasized ; in the first place, they are
not always clearly understood, but are seen only through a
veil of general principles[2], and secondly, the last part, about

---

[1] The physicists have no word that fits exactly. We use the term
" Coherence " from the theory of radiation, as being the least in-
appropriate.

[2] E. Wasmann, e.g. *Die psychischen Fahigkeiten der Ameisen*, 2nd ed.,
1909, p. 108, seqq. has sharply defined this contrast. But he absolutely
denies intelligence in animals, and further points to a logical theory of
intelligent conduct (intelligence) in the case of man, which I cannot
accept. O. Selz, *Die Gesetze des geordneten Denkverlaufs, I.*, 1913, treats
of reproductive thought in man from a point of view somewhat related
to mine.

insight, appears to some students not at all obvious, but rather as a sort of belief in miracles. No such superstition is meant or prepared here, and nothing that has been said involves it in the slightest.]

How one is to explain that the field as a whole, the relations of the parts of the situation to one another, etc., determine the solution, belong to the theory. Here we have only to exclude the idea that the behaviour of the animals is to be explained by the assumption according to which the solution will be accomplished without regard to the structure of the situation, as a sequence of chance parts, that is to say, without intelligence.

In the description of these experiments it should have been apparent enough that what is lacking for this explanation is that most necessary thing, a composition of the solutions out of chance parts. It is certainly not a characteristic of the chimpanzee, when he is brought into an experimental situation, to make any chance movements, out of which, among other things, a non-genuine solution could arise. He is very seldom seen to attempt anything which would have to be considered accidental in relation to the situation (excepting, of course, if his interest is turned away from the objective to other things). As long as his efforts are directed to the objective, all distinguishable stages of his behaviour (as with human beings in similar situations), tend to appear as complete attempts at solutions, *none* of which appears as the product of accidentally arrayed parts. This is true, most of all, of the solution which is finally successful. Certainly, it often follows upon a period of perplexity or quiet (often a period of survey), but in real and convincing cases, the solution never appears in a disorder of blind impulses. It is one continuous smooth action, which can be resolved into parts *only by abstract thinking* by the on-looker ; in *reality* they do *not* appear independently. But that in so many " genuine " cases as have been described, these solutions *as wholes* should have arisen from mere chance, is an entirely inadmissible supposition, which the theory cannot allow without renouncing what is considered its chief merit.

[I have noticed from myself and others, that what is particularly enlightening as to the ape's behaviour are the pauses mentioned above. A local colleague convinced, like most students, of the general value of the chance theory for animal psychology, came to see the anthropoids. I chose Sultan for the demonstration. He made one attempt at solution, then a second, and a third ; but nothing made so great an impression on the visitor as the pause after that, during which Sultan slowly scratched his head and moved nothing but his eyes and his head gently, while he most carefully eyed the whole situation.]

This sort of question can best be answered by actually observing the facts with which this theory asserts it can explain *all* our experiments, thus making oneself, by observation, more capable of judgment. Behaviour fit for such an examination occurred in the building-with-boxes tests. Here, in the solution which, taken *en bloc*, " higher box up," was quite clear, the final result was achieved only by chance, and after an almost entirely unintelligent muddling around. This happened so often, and so uniformly with all the animals experimented on, that I can claim to know exactly the procedure asserted by that theory to be general. *It should, therefore, be the more emphasized that the most striking difference exists between this conduct, obviously ruled by chance, and the behaviour described as " genuine " in clear solutions.* In addition, the descriptions of these experiments have shown how unwillingly the chimpanzee embarks on procedure, of which the general outline will come to him as a genuine solution, but whose more detailed execution he must attempt as mere trying, that is, leaving it to chance. The animals

would never have hit on this kind of trying, if an attempt
genuine in outline had not put them in a position, with the
special conditions of which they were not able to deal. The
fact that the animals on such an occasion do make blind
movements does not in any way contradict the assertion
that, as a rule, and in reasonable testing conditions[1], disorder
of impulses is not observed.

In these experiments, whenever chance may have effected
or favoured a solution, the fact is always mentioned. In
complicated experimental conditions (see the next chapter,
such cases are more frequent, but it must be said from the
beginning that even then the course of the experiment does
not entirely agree with that theoretical interpretation. In
the first place, it may happen that the animal will attempt
a solution which, while it may not result in success, yet has
some meaning in regard to the situation. " Trying around "
then consists in attempts at solution in the *half-understood*
situation ; and the real solution may easily arise by some
chance outcome of it, i.e. it will not arise from chance impulses,
but from actions, which, because they are *au fond* sensible,
are great aids to chance. Secondly, a lucky accident may
occur in some action, which has nothing to do with the objective. Here again, there is no question of a meaningless
impulse—the chimpanzee only gives way to these, as already
remarked, when driven to it—but of some kind of intelligent
activity, even if with no reference to the objective. This
is what probably occurs, when Sultan discovers the way to
combine two sticks ; only a Philistine would call his playing
with these sticks " meaningless impulses," because it follows
no practical purpose. That an accident helped him is not
the most important fact in either case ; the important thing

[1] Arrangements are usually made—in accordance with the requirements of the problem—so that it is not easy for accidental solutions
to take place.

is how the experiment then proceeds. For we know from Man that even an accident may lead to *intelligent* further work (or intelligent repetition), especially in scientific discoveries (compare Oerstedt : *Current and Magnet.* Thus Sultan's behaviour, when he has once carried out his usual play, " put stick in hole," with both the bamboo rods, is exactly the same as if he had discovered the new procedure in a genuine solution. After this there is no doubt that he makes use of the double-stick technique intelligently, and the accident seems merely to have acted as an aid—fairly strong it is true—which led at once to " insight."

If one does not watch attentively, the crude stupidities of the animals, already referred to several times, might be taken as proofs that the chimpanzee does, after all, perform senseless actions, a sequence of which may, by chance, give rise to apparent solutions.

The chimpanzee commits three kinds of errors :—

1. " *Good errors,*" of which more will be said later. In these, the animal does not make a stupid, but rather an almost favourable impression, if only the observer can get right away from preoccupation with human achievements, and concentrate only on the nature of the behaviour observed.

2. *Errors caused by complete lack of comprehension of the conditions of the task.* This can be seen when the animals, in putting a box higher up, will take it from a statically good position and put it into a bad one. The impression one gets in such cases is that of a certain innocent limitation.

3. *Crude stupidities arising from habit* in situations which the animal ought to be able to survey (e.g. dragging the box to the railings—Sultan). Such behaviour is extremely annoying—it almost makes one angry.

Here we are dealing with the third class, and it is easily seen that these mistakes are not at all liable to confirm the

chance theory. This kind of behaviour never arises unless
a similar procedure often took place beforehand as a real
and genuine solution. The stupidities are not accidental
" natural " fractions, from which *primarily* apparent solutions
can arise—I know of no case in which such an interpretation
is even possible—they are the *after-effects* of former genuine
solutions, which were often repeated, and so developed a
tendency to appear *secondarily* in later experiments, without
much consideration for the special situation. The preceding
conditions for such mistakes seem to be drowsiness, exhaustion, colds, or even excitement. For instance, a chimpanzee,
when he performs an experiment for the first time and cannot
reach the objective lying outside the bars without an implement, will never have the " accidental impulse " to drag
a box to the bars, and even get up on it. On the other hand,
one may see that actually, after frequent repetition of a
solution originally arrived at genuinely, and in the consequent
mechanization of the proceeding, such stupidities are easily
committed. Not infrequently have I demonstrated an
experiment to interested observers, and, for the sake of
simplicity, usually chose the opening of a door, in front of
the hinge side of which the objective was hanging. After the
animal had done this about twenty times since the first
solution, and always at the same place, there began to appear
a tendency to fetch down objectives hung high up with the help
of a door, even when other methods were more obvious, and
the use of a door had been made very difficult, in fact, almost
impossible. And if attempts at other solutions developed,
they were more or less under the influence, or magnetic
power, of the door. Chica, for instance, made out of the
jumping-stick method, which she had in its simple form
completely mastered, a combination of this and the door-
method ; and quite unnecessarily, because it was by no
means an improvement. Before the door had come into

intelligent use for the first time, the chimpanzees had paid
no attention to it in any experiment, not even when the experiment took place opposite to it.

According to this, processes, originally very valuable,
have a disagreeable tendency to sink to a lower rank with
constant repetition. This *secondary self-training* is usually
supposed to bring about a great saving, and it may be so,
both in man and in anthropoid apes. But one must never
forget what a startling resemblance there is between these
crude stupidities of the chimpanzees arising from habit,
and certain empty and meaningless repetitions of moral,
political, and other principles in men. Once all these meant
more, one cared about the " solution " in a predicament deeply
felt or much thought about ; but later the situation does not
matter so much, and the statement of the principle becomes
a cliché.

It should now be clear enough that these meaningless
reproductions of originally genuine and correct solutions
have absolutely nothing to do with the accidental and confused
production of " natural " impulses of the theory discussed
above.

[For the rest, it will be best simply to give the whole list of
these stupidities :

1. Sultan puts one box on top of another, where the objective
was before, not where it is hanging now ; the animal is quite
exhausted (8.2.1914).

2. Sultan drags a box to that spot at the bars, opposite
which the objective is lying (outside), and turns first one
side, then the other towards the bars quite stupidly ; fetches
more boxes, and begins as if to build. The animal has been
performing experiments with boxes for about four weeks
continuously ; the experimenter is partly to blame (19.2.1914).

3. In the same experiment Sultan draws the observer
thither, and climbs on his back, as if the objective were hung

high up ; he jumps down again at once, and then follows the solution described above, p. 148 (19.2.1914).

4. Sultan drags a box to the bars, where the objective is lying outside them (20.4.1914).

5. Grande commits the same stupidity (14.5.1914).

6. With the objective outside the bars and at some distance, Grande drags stones about in her cage, as an after-effect of repeated experiments in the same place, in which stones served as a footstool for her (19.6.1914).

7. Koko shoves the box in the direction of distant fruit, and for a moment does not use it as a stick, as on the day before, but as a stool. The animal is very excited (1.8.1914).

8. Koko does the same when in a fury (6.8.1914).

A similar thing is hinted at once when, with the objective hanging high up, Sultan goes towards the nearest door—a good three metres away—takes hold of it, but lets it go again after a look at the objective, and turns to other methods. In this case he is near a meaningless reproduction, but is prevented by a survey of the requirements of the situation (13.3.1916).

Rana's habit of beginning over and over again to jump with tiny sticks, is hardly to be included here. Rana's brain, as it were, runs away with her, and, of course, it would be fine if she could jump like that. The animal will indicate the same kind of behaviour, even when she sees clearly that its execution is impossible.

That is all ; practically all the cases have already been mentioned in the reports of the experiments. No one will assert that they represent the main character of our observations.

The chief cause of these phenomena (mechanization) need not, necessarily, according to the above, lead to externally observable effects of the nature of crude stupidities. Every solution repeated often under the same circumstances, and adequate to them, changes somewhat in nature, and perhaps finally will not be so intelligent even in this, its original *milieu*, though still adequate. I must say that I like the behaviour of the chimpanzees during their tenth or eleventh repetition of a solution less than that in the first or second. Something is spoilt in the chimpanzee even when many different experiments follow each other in quick succession, but particularly when the same ones are repeated. I may not always have sufficiently considered this possibility in my eagerness to make researches.

The facts we are speaking of, by the way, seem to represent almost a reversal of what the theory we have discussed regards as the effect of repetitions. According to it, procedure developed by accident becomes smoother through practice, and more like a genuine solution. This may be true, where the theory applies ; the chimpanzee's *genuine* solutions, at any rate, do not become more valuable in themselves through constant repetition, even if they appear more quickly, etc.]

For one who has actually watched the experiments, discussions like the above have something comic about them. For instance, when one has seen for oneself, how in the first experiment of her life (cf. above, p. 62), it did not dawn on Tschego for hours to push the obstructing box out of the way, how she merely stretched out her arm uselessly, or else sat down quietly, but then, fearing the loss of her food, suddenly seized the obstacle, and pushed it to one side, thus solving the task in a second—when one has watched that, then to " secure these facts against misinterpretation " seems almost pedantic. But the living impression will not be reproduced, and many a question can be raised on the words of a report, which would not even occur to anyone after some observation. Nevertheless, it may be that after these discussions, the description of a further experiment carried out as a model will be particularly instructive ; an experiment which is characterized both by its simplicity and its unequivocal relation to several theories.

A heavy box is standing upright at some distance on the other side of the oft-mentioned bars ; one end of a stout string is affixed to it, and the string itself is laid down obliquely so that its free end lies between the vertical bars of the railings. Half-way between the box and the bars fruit is tied to the string (cf. Fig. 14) ; it cannot be reached from the bars as it is, but only if the string is laid straight. (19.6.1914) First of all, Chica pulls in the direction in which the string is

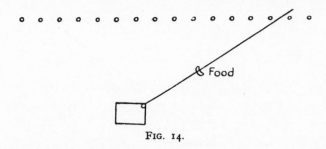

Food

FIG. 14.

lying, and so hard that the board of the box breaks, the string is freed, and the objective can be pulled to her. The box is then replaced by a heavy stone and the string tied round it. As the simple solution by pulling is no longer possible, Chica takes the string in one hand, passes it round the bar to her other, which she puts through the next space, and so on, passing it thus until the string is at right angles to the bars, and the objective can be seized.

Grande seems at first not to see the string, which is grey and lying on a grey ground. She drags stones about sense-lessly (compare p. 197)—an after-effect of earlier experiments —tries to detach an iron rod from the wall, which she presum-ably wants to use as a stick, and at last sees the string. After this the experiment runs as with Chica, a solution without any hesitations.

Rana first pulls twice in the direction of the string, then suddenly changes the direction completely, while trying to pull the string to a spot just opposite the one at which it is tied (cf. Fig. 15) ; at the same time she stands opposite this point herself and keeps on looking at the objective and pulling the string parallel to the bars. This vain attempt is made twice in succession, in separate stages, and then is replaced by the proper solution, as in Grande's and Chica's cases. This experiment shows that the task consists of two parts : one, crude in its geometric and dynamic properties,

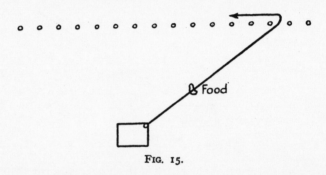

FIG. 15.

" turn string at right angles to bars so that the objective comes nearer," and the more refined special problem, arising from the structure of the bars. Chica and Grande solve both parts at once ; Rana solves the first one quickly, and the second one only later.

Sultan pulls for a moment like Rana (cf. Fig. 15) and immediately afterwards solves the problem completely, like the others. It becomes quite clear through this that the crude dynamic problem can be solved without any regard for the special one (the second problem), which in this case only seems to be noticed through non-success. Similar effects were encountered in the building with boxes.

Tercera cannot be cajoled into taking part. Tschego and Konsul show—in case it is not already realized—that the solution is not obvious ; for neither gets any further than pulling the thread in the direction in which it is lying.

(21.6) The experiment is repeated with Chica, but this time the thread lies on the floor *turned in the other direction.* The animal does not pull at all in the direction of the thread, but starts the hand-over-hand process straight away, in the *opposite* direction from the previous experiment, till the goal can be reached. After this I did not think it necessary to make the same experiment with the other animals.

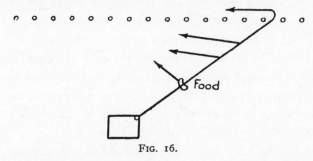

FIG. 16.

After the foregoing explanations, it need hardly be pointed out again that experiments like the one just described give *better* information about the chimpanzee than the usual animal tests with complicated locked doors, etc. Also it should be realized that an experiment so simple and clear as this contain the whole problem to be considered.

If anyone should still be of the opinion that such simple solutions are obvious and have nothing to do with intelligence, I can only invite him to show definitely and exactly the way in which the procedure comes into being. I am afraid no psychologist is able to accomplish this at present.

I separated the two parts of the problem, which, as we saw, are independent, and I shall now consider only the crude one and its solution. This can be characterized simply by the sketch (cf. Fig. 16), leaving out of consideration how the animal, at the first moment of the solution, actually performs the arrow movement in detail (taking the bars into consideration or not).

Do the animals arrive at the solution in accordance with the theory we have discussed ? If so, we should expect to find in all cases the appearance of a large number of impulses which might, in some of the chimpanzees, perhaps, accidentally contain the right " fragments " in the right succession. In reality Grande is the only animal that does anything senseless, and that in the form of a habit stupidity, when she had not yet thoroughly surveyed the possibilities of the problem ; when she sees the string, a new stage of behaviour sets in, and immediately afterwards a perfectly clear solution is achieved. Altogether only two movements (" impulses " may occur in lizards, but rarely in chimpanzees) really take place with regard to the objective. These two movements are :—

1. *Pulling in the direction of the string*, i.e. a sensible proceeding, the practicability of which Chica once proves. No man, still less a chimpanzee, can otherwise find out if the string will not really come loose from box or stone.

2. Pulling at the string, or continuous passing of the rope hand-over-hand—in both cases in the *right direction for a solution* (see arrows in the diagram).

Not in a single animal was anything approaching a direction midway between these two observed, much less a third quite new one, etc. Where the more primitive tendency appeared first (in the direction of the string), the jump to the other one was yet made quite abruptly.

I should imagine that everybody must feel that we have here a very clear, though peculiar occurrence, and one which has nothing at all to do with the postulates of that theory. Are we to squeeze and force the facts to make them fit in with that theory, just to suit the so-called principle of scientific economy ? In this case the observer is forced to the conclusion that attempts 1 and 2, appearing as wholes, yet each on its own, are a direct result of a visual survey of the situation. A certain scientific attitude, which one might also formulate as a principle, the " principle of maximum scientific fertility," would lead one to begin the theoretical considerations with this character of the observations, and not to eliminate it at whatever cost as the theory of chance does.

There would be no need to discuss this theory further if the previous life of the chimpanzees we have tested were known to us in all details, from birth to the moment of testing. But unfortunately this is not the case ; and even if the possibility that in the *experiment* the solutions arise by chance be excluded, yet the possibility remains that they were developed *before*, within the scope of the theory, by accident, that they were repeated and improved, and now seem to appear as genuine solutions.

It is always difficult to contend against arguments which are outside the domain of possible proof. In this case, however, not even the overstepping of the bounds of experience will be a weakness in the argument, for naturally the chimpanzees we tested had passed several years as lively animals, uncontrolled, in the jungle of the West Coast, and, while there, came into contact with several objects similar to those employed in some of our experiments. Thus it becomes necessary to consider whether this circumstance does not influence the significance and the factual value of the experiments.

But two points must be kept rigidly in mind, if the object of the discussion is not to be lost :—

1. The fact that the animals have had to deal with single objects or situations prior to the experiments, has not necessarily any direct connexion with out problem. It is only when, exactly according to the theory, during this previous period, meaningless but successful chains of actions, externally like the behaviour observed here, have been formed—acci-

dentally, and selected by success—that " previous experience " speaks against the value of these experiments. I am far from asserting that the animals tested in the second chapter have never had a stick, or anything like that, in their hands before the experiment. On the contrary, I take it for granted that every chimpanzee above a certain very low age has had some such experience ; he will have seized a branch in play, scratched on the ground with it, and so on. Exactly the same thing is very frequently observed in small children of less than a year, so that these, too, had their " experience " with sticks, before they used them as implements to pull things towards them that they could not otherwise reach. But just as this does not at all prove that they get accustomed to the use of implements in the mere play of chance and quite without insight, reproducing it again without insight at two, four, or twenty years of age, so also it does not follow for the chimpanzee, whose test-stick is not the first he has ever had in his hand.

2. I am by no means trying in this work to prove that the chimpanzee is a marvel of intelligence[1] : on the contrary, the narrow limit of his powers (as compared to man's) has often been demonstrated. All that has to be decided is whether *any* of his actions have ever the characteristics of insight, and the answer to this question of principle is at present far more important than an exact determination of *degrees* of intelligence. On the other hand, the theory of chance, discussed here as a general principle of interpretation has no interest in the mere *diminution* of the *number* of intelligent acts in experiment, but, in order to be convincing, the theory must explain *all* tests, without exception, consistently with itself. And it fails when, even though some results observed are explained by it, others are not. In the latter

---

[1] One is obliged to-day to mention in a serious book that the chimpanzee has up till now shown no inclination or gift, for instance, for the study of fourth roots or elliptic functions.

case, when the general application collapses, there will be less temptation to explain certain kinds of behaviour as products of accident, which, by their nature, do not invite such an interpretation, though they may be forced under this theory.

The past history of these animals, before the tests, is not altogether unknown. Since at least the beginning of the year 1913, they have been carefully watched, and for a further six months before that date, we can rest assured that any practice in a number of test-situations was impossible, because the animals were confined in the narrowest cages, with no " objects " in them (in Cameroon, on the voyage, in Tenerife). According to the information of my predecessor, E. Teuber, during the year of observation *before* these tests, Sultan and Rana did not get beyond using ordinary sticks (without any complications) for lengthening of the arm, and jumping—the others did not even achieve this much ; occasional throwing of stones was observed, and in one case the fabrication of an implement as described above (pp. 101-2), when Sultan takes the shoe-cleaner to pieces.

In any case, the following circumstance is important : when it is a question of the principal decision, whether insight occurs or not, then for any explanation to be in accordance with the chance theory, *not the slightest trace of insight* must occur, not in the most hidden, or in the most innocent, disguise. Therefore since everything, to the smallest details, was to be put together out of chance combinations of elements, and rehearsed, until it could *seem* to appear as a single and intelligent action in the experiments, so we shall, in general, have to assume, not *one sole* former occasion in a similar situation, but a *series of repetitions* of such occasions. Only then somebody might say with conviction that this procedure or that, or rather *all* the lines of action here observed have had their origin and development, in accordance with the principles of the theory.

I remarked above that the general principles of higher psychology often had a tendency to hide rather than to clarify the things to be explained. For instance, when we say that the objectively-useful employment of a stick, as a means of reaching otherwise inaccessible objects, developed by accident and the selective working of success, it will sound very precise and satisfactory. When we look closer, however, our satisfaction with the general principle is soon diminished, if we are really serious in making the condition " without a trace of insight." Let us assume, for instance, that the animal seized a little stick *by accident* at a time when some food, otherwise unattainable, lay at some distance. As, for the ape, the stick and the objective have nothing to do with each other, we have to ascribe it to chance also, if, among a large number of other possibilities, the animal brings the stick into the vicinity of the object desired. For, of course, we must not assume that this action occurs all at once, as one. With one of its ends in the neighbourhood of the objective, the stick has still nothing at all to do with the objective, as far as the animal is concerned, he " does not know " that he has arrived objectively a little nearer to the attainment of the goal. The stick may be dropped, or pulled back, or pointed in all the directions of a sphere with the animal as centre ; and chance will now have to work hard until from all the possibilities *one* emerges, namely that the end of the stick is put down behind the objective. But again, this position of the stick tells the unintelligent animal nothing ; as before, the most various " impulses " may appear and chance might well have reached the limit of its capacity, if the animal now makes an accidental movement which brings the goal a little nearer to it. But this again the animal does not understand as an improvement of the situation ; for it understands nothing at all, and poor, exhausted chance, which has to do all the work that the animal itself is unable to do directly, must now prevent the stick from being dropped, drawn back, and so forth, and must bring it about that the animal keeps the right direction in further chance impulses. It may be said that there are very various sequences or combinations of impulses containing, for instance, as their last constituents "stick behind objective," and after that "the objectively fitting impulse." That is correct, and the possibilities open to chance, if it is to do this great work, become thereupon more numerous. And yet even now nothing is spared to it ; for the majority of these combinations contain, of course, factors objectively quite meaningless, which only follow upon each other in such a way that the whole series *finally leads to the two elements mentioned above.* Therefore, if the first favourable combinations, of which these elements form the end, contain such objectively-meaningless components, chance must later complete the work by means of a large number of other favourable cases, until a perfectly smooth, and *seemingly intelligent*, procedure matures with the help of the (at first, probably extremely rare) successes ; for as the use of the stick is observed here for the first time, it contains in no case a thoroughly false component, even if (as with Koko) weakness of the arm and clumsiness act as somewhat of a hindrance.

At this juncture it will probably be objected that the desire for the objective, the general *urge of the instinct* in its direction, is being left out of consideration. To this we reply : in the first place, to conform to the theory, we assume that this " instinct " is perfectly blind, that *the animal is not in any way aware that he is nearing its goal* by taking this direction— for otherwise the theory would be untrue to itself ; secondly, according to the theory, this instinct exists *for the body of the animal,* and for the innervations of his limbs, *not for the stick* he happens to hold in his hand. I want to know therefore : if the animal, following that impulse, moves his arm in the direction of the objective in order to catch hold of it, why should he keep the stick, of which his instinct knows nothing, in his hand, rather than open his hand to seize the objective, as at other times, and thus let go of the stick ? For, all this time, the stick has, in the animal's eyes, *nothing to do with the objective*. Should he, however, contrary to this demand of the chance theory, continue holding the stick in his hand, that would, with his lack of any trace of insight, be possible in a variety of very different ways. It may be held right in the centre, so that the stick is parallel to his front and sideways, or it may be grasped at the extreme end, the other end pointing back towards the animal, upwards to the sky, or down to the ground, etc. For if nothing is assumed but the impulse of instinct in the direction of the objective, and accidental movements—*intelligence*, to the contrary, remaining wholly *excluded*, one way of holding the stick is as good as another and the different possibilities are limited only by the animal's muscular power ; because success will have its selecting effect at the earliest after *one* favourable combination only. And so chance which has already, in opposition to the theory, left the stick in the animal's hand, has still plenty to do before it succeeds in obtaining the right manner of holding the stick, in eliminating the false elements by the help of chance *successes*, and obtaining a mode of procedure, superficially similar to intelligent behaviour.

[It might be further objected that it is not necessary that the first result be obtained *all at once* in an action of this kind : all kinds of *subordinate actions* might first develop accidentally, and be later more easily united in a combined action. But this reflection does not help us either, because the permutations corresponding to the said subordinate actions might occur once, but they cannot be co-ordinated into firmly connected partial actions, which alone would be of use. This is because *that " success " is lacking*, which, according to the theory, ensures the connexion of the individual impulses. What does it matter if the animal is once in a way led by accident to putting one end of the stick behind the objective ? That is no success, within the meaning of the theory, so long as, in accordance with the theory, the animal is without any trace of understanding. And so chance must either take one more step and continue its permutations to the end until objective and animal meet, or else it immediately thereafter strays into quite irrelevant impulses ; and then, according to the theory, there is no tendency ever to repeat this combination, the last lap of which is—" point of stick behind objective."]

This theory of chance is preferred to many other attempts at explanation, because it is held to be specially exact, and in exceptionally good correspondence with the demands on scientific thought. For this reason many people would no doubt like to see not only the use of the stick, but *all* other performances explained by it as above described. Little as there is to be said against the theory, in cases in which success can be easily achieved by chance (as, for instance, when an animal locked in a small box, tries blindly to get out, and in the course of its disordered movements, happens to push a lever which opens the door), there does not seem to be much value in it, just from the scientific point of view, when it is used to explain such experiments as the ones described here.

Those scientific ideas with which we here come into conflict are the ones which suggested to Boltzmann the (as yet) most comprehensive and important formulation ever made of the second law of thermo-dynamic. According to this law, neither physics nor theoretical chemistry allow of the fortuitous formation of a well-directed *total* movement in the course of the permutations of a large number of small chance movements, which are mutually independent, irregular and all of them equally possible.[1] For instance, in the case of Brown's molecu-

---

[1] In physics one speaks better and more exactly about " *Zustandselemente* ". It would lead too far to go into this more closely. C.f for example, Planck, *Acht Vorlesungen über theoretische Physik*, 1910.

lar movement, it is impossible for a suspended particle, pushed hither and thither fortuitously and irregularly, to be suddenly projected one decimetre in a straight direction. If such a thing did happen, without doubt a source of error would have entered, i.e. *an influence not following the laws of probability*. Now, whether it be a question of Brown's molecular movement, or of the so-called chance impulses of a chimpanzee, makes no essential difference here ; for the *bases* of the second law (according to Boltzmann) are of so general a nature and so obviously valid for more than thermo-dynamics (namely, for the whole domain of chance) that they are applicable also to our (alleged) subject-matter, the "impulses". Anyone who reproaches us for playing with analogies must surely have misunderstood the fundamental thought of Boltzmann (and Planck). There does, however, exist a quantitative difference between the thermo-dynamic case and our own. In what degree the appearance of a special combination is improbable (up to practically quite impossible) depends on the number or size of the independent elements which are combined. One easily sees that in this respect the " impossibilities " of thermo-dynamics are not quite achieved by those of the animal tests (as above described in the use of the stick), as we have to deal here with *less* members (i.e. possible "impulses") and these, compared with the total proceeding, are *relatively still great*. Of course, this alters nothing as regards the direction of our reflections and the fundamental doubtfulness of the assertions of the theory so long as one *does exclude non-accidental forces*, and thoroughly considers individual cases, with their immense demands, as was done in the stick experiment.

[Neither the general direction of the "instinct-impulse" towards the objective, nor the further development in "selection through success", alter anything in the unfavourable condition of affairs ; the former does not, for the above-given reason, the latter because from the first it presupposes the right succession of lucky hits, which is not likely to occur, but without which " success " will have no opportunity to work at all.]

Since similar considerations have been advanced on questions dealing with evolution by Bergson and E. v. Hartmann, and as they play a great part in vitalistic literature, the following observations seem relevant. E. v. Hartmann considers it impossible that the bird arrives at its nest by chance, and he concludes that " the unconscious " is the builder; Bergson considers accidental arranging of the elements of an eye altogether too improbable, and, therefore, makes the " *elan vital* " accomplish the miracle. The neo-vitalists and psycho-vitalists, too, are equally unsatisfied with Darwinian chance, and they consider that everywhere in the specifically animate realm " forces with a purpose," of the general type of human thinking, are required for explanation, though these forces are not actually experienced, as thinking is. *The only connexion that this book has with such a line of thought, is that here, too, a theory of chance is rejected.* But generally the transition from the rejection of this chance theory to the acceptance of one of those doctrines is regarded as almost obligatory, and I, therefore, wish to emphasize that the alternative is not at all between chance and factors outside experience. In that opposition lies the fundamental error, that all that happens in inanimate nature is to be considered as subject to the laws of probability, whereas, after all, great parts of physics have nothing to do with chance. As certain as it is that the study of physics does not consist solely of the laws of irregular heat movement, even so certain is it that one need not from a view like the above, which contradicts the theory of chance, jump to the assumption of agents outside experience. It seems particularly surprising from the stand-point of physics that one should continually insist on speaking here of "either—or", when after all there are quite other possibilities[1].

I think I have shown that the theory of chance can in no way be considered exact in every case, and that in such performances as those described here, absurd demands are made of chance, whereas natural science in particular does not allow such blind confidence beyond certain limits. It is, therefore, advisable to throw another glance at the experiments themselves.

According to the chance theory, we never have before us a first occurrence, but—as the proceeding (compare use of stick) measured by "impulses" is relatively complex and yet perfectly smooth—always a case which is the product of frequent repetition. The objectively correct use of a box, etc., as a footstool, for example, could not be developed in less than, say, fifty repetitions. At least as often as that, an objective placed high up ought to have been unattainable by the animals, in any easier manner, and, at the same time, a box or similar object should have been at hand. One has only to remember how improbable it is even, that the animal in such a situation should seize the tool at all or move it, *as long as it has no insight whatever*. The more thoroughly one studies this case— as is much to be desired—the more will one consider far higher minimum numbers of repetitions necessary to chance in the development of the behaviour.

The same closer consideration of individual cases shows also that generally not even the simplest preliminary condition for such an expensive play of chance is present. How often, in any case, could a chimpanzee, under his normal conditions of life, be in a position to need, for example, a footstool to reach a high object—presumably the fruit on a tree ? Over and above any use of implements, the solution for the chimpanzee

---

[1] Compare my book *Die Physischen Gestalten in Ruhe und im stationären Zustand*, Brunswick, 1920.

is the roundabout way *in its literal sense* ; every time the detour seems even a remote possibility to us (and beyond this human limit, too) these animals do not hesitate at all, and certainly show no other " impulses " ; they *start on the round-about way immediately.* At the beginning of the experiments it was my special task to make this easy proceeding impossible for them (compare above, p. 11 seqq.). If it is a question of trees—and where else in the Cameroon jungles could anything be hanging high up ?—I maintain that there is scarcely anything that the chimpanzee cannot reach in some roundabout way or other. One must have seen for oneself how even a chimpanzee who is a bad gymnast (as, for instance, Sultan, whom I sometimes took out of doors) jumps from one tree to another, seems to fall, slips, and so forth ; how he falls into the thin foliage of a tree which has no proper branches, but only leaves and the tiniest twigs, and yet, catching hold of it for the fraction of a second, lets it act as a sufficient brake to enable him to continue swinging himself along, jumping, and falling, until he comes to a standstill again in any firm spot he chooses. Thus I must emphatically deny (in the case of the use of boxes) that the animals had enough occasion to be forced, through the exclusion of the roundabout gymnastic methods natural to them, to combine any number of other impulses. Under natural conditions they arrive anywhere without stools, and only Man, making experiments, brings them into situations where such roundabout methods are excluded objectively, or through Man's prohibitions. The same may be said about the use of a stick for pulling objects to them which would otherwise be unattainable (in Tenerife such an action was not noticed in Tschego before the experiment, and Nueva and Koko were examined immediately on their arrival), about the pushing away of a box which is in the way of the railings (when at liberty the chimpanzee, of course, takes roundabout paths round stone blocks or thick tree-trunks). As a number of farther experiments presuppose an appropriate employment of stick and box, they too lack— and for the same reason—the previous history necessary for adequate combination.

[Once again let me repeat : a number of objects become familiar to the animals in some way or another before the experiments. But there is a big difference between touching a stick and using it " intelligently ". If one now abandons the theory, and asks whether Nueva, for instance, has not sometimes *in intelligent play* pushed a stone about with the stick, the answer is doubtful, in such a changed aspect of the problem. For even with little insight many things, of course, become easy which could never occur by accident. I am very much inclined to answer the question in the affirmative, as every day, during the play of the animals, when they understand very well what they are doing, such things do happen.

Should any doubts remain about the foregoing, none can be entertained about my assertion that, in some cases, the situation either faces the animal *for the first time* when he performs the experiment, or else he may have experienced something similar, but only very rarely. The model test described above (p. 199) is an example. Who can seriously assert that any of the animals had been in a similar situation prior to this test, in which they are confined to a space behind railings, a string fastened outside lying obliquely on the ground, approximately in the middle of which a piece of fruit is tied, so that only a certain turn of the rope will make the object attainable ? Even if the test is simple, the animals will not have experienced such a thing, and yet four of them, independently of each other, solve the principal task all of a sudden. Never, before the experiment, did an object hang in front of a door, and yet, in such an experimental situation, the door is suddenly looked at sharply, and at once opened, a clear solution. Leaving aside the question of how Sultan arrives at the use of a box at all, there remains the other : what makes him, when the test is performed, take out the encumbering stones ? Where did he ever get the chance of making blind combinations in a situation like this ? And, further, there can only have been very few single cases, not by any means as many as required by the theory, of Nueva not being able to reach the objective with a stick too short and, by chance, finding a longer one close by, with which she could reach the first one, etc.—of course always through chance impulses.

It is indeed a great effort to argue so much against an explanation for which the observations give no grounds. Finally, I shall once more call attention to the character of these observations, which tell one more than any argument could, contrasting them with the requirements of the theory.

1. The animals are supposed to have accidentally got accustomed to such solutions in previous life ; an *extremely familiar* action, the result of very much practice, was, it is presumed, observed, which, on account of its extreme familiarity, looks exactly like an intelligent solution. But the best and most obvious solutions which I observed, often occurred suddenly, after the animal had been quite helpless at the beginning of the experiment, and sometimes for hours after. Whoever considers Tschego's first experiment (when the box was in the way at the bars) or Koko's (use of box as a stool) to be the repetition of long-practised, mechanical and meaningless products of habit, does so certainly in opposition to the impression which observation of the procedure must make.

2. The animals are supposed to have so developed, strengthened, and perfected their performance, through the selection by success of "impulses", that they can now "easily" reproduce it *in this form*. No single experiment fulfils this requirement, as practically none is performed twice over in the same way ; indeed, the movements by which one single one is performed vary greatly   The door is opened from the

ground, but also the animal sitting on top of it ; when the box is standing in the way, it is pushed away by a corner from the bars, or thrown back over the bottom edge. If the box is to be brought underneath the objective, the same animal will drag, carry, or roll it along, just as the mood takes him. *The only limit is the sense of the proceeding.* For this reason no observer, even with the best of efforts, can say : " the animal contracts such or such a muscle, carries out this or that impulse ". *This would be to accentuate an inessential side-issue, which may change from one case to another.* To give the essentials, it is necessary to use expressions in describing all this, which themselves involve meaningful actions ; for instance, " *the animal removed from the bars the box which stood in the way* ". Which muscles carry out which actions is entirely immaterial.

3. There are other variations not so unimportant, which likewise run counter to the theory, *but arise directly through unforeseen circumstances*, and represent the answer to these circumstances ; these *cannot possibly all have been rehearsed.* The animal then does not continue carrying out a rehearsed programme meaninglessly, but answers to a disturbance by a corresponding variation. This is often the case when using the stick. It is easy to say, that the animal fetches an object with the stick, but in reality it does so each time in a different way, because on uneven ground each movement brings the object into a different position which requires special handling. When Sultan for the first time pulled one stick towards himself with another, the test went very smoothly on favourable ground. But the next time the stick encountered a pebble while he was drawing it to him, and so he could not get it any farther, as it was turned round and pointed straight towards him (lying lengthwise). The animal stopped *at once*, first pushed the stick, with careful little pokes of the second stick, crosswise again, and then pulled

it to him. One may truthfully say that in the majority of
cases when the stick is used, the solving of the chief problem
brings in its course small, unforeseen additional problems,
and that, as a rule, the chimpanzee immediately makes the
necessary modifications. Of course there are limits here, too
—they will be dealt with in the next chapter—but we are not
asserting that the chimpanzee can do as much as an adult
man. On the other hand, it would be simply nonsense to
assert that the animal has gone through special combinations
of accidental impulses for all these different cases and variations.

4. *Success* is supposed to have selected and joined together
the objectively suitable combinations of impulses out of all
those that occurred. But the animals produce complete
methods of solution, quite suddenly, and *as complete wholes
which may, in a certain sense, be absolutely appropriate to the
situation, and yet cannot be carried out. They can never have
had any success with them*, and, therefore, such methods were
certainly never practised formerly (as they would have to be,
according to the theory). I would remind the reader how
two animals suddenly lift a box that stands too low, and hold
it high against the wall; how several of them endeavour to
stand the box diagonally so as to make it reach higher; how
Rana joins two sticks that are too small, to make one that
looks twice as long; how Sultan steers one stick with another
from quite a distance right up to the objective and thus
"reaches" it, so to speak. In the second part of this investigation a particularly curious case will be described, in
which several animals, when a block of stone prevented them
from opening a heavy door, *suddenly* made the greatest efforts
to lift the heavy door over the stone. How can selection by
success have trained them to such " good errors " ?

After all this, as far as I can see, even an adherent of the
chance theory must recognize that the reports of experiments

here given do not support his explanation. The more he tries to advance more valuable data than the general scheme of his theory, and really think out and show how he would explain and interpret all the experiments in detail, the more will he realize that he is attempting something impossible. Only he must keep in view the condition that not even in the most innocent form or in the smallest detail is intelligence to be allowed to co-operate as insight into the structure of the situation.

Whoever is not sure from the very beginning (as a disciple of scientific economy) that this theory only and no other may be applied to animals, must be asked once again to look through the reports of some of the experiments. Even if that will give him but a faint idea of what direct observation of the actual occurrences teaches one—that cannot be adequately reproduced—he may perhaps feel that, besides the theory, such extended discussions about it are not suitable here ; *to such an extent do observations, and the manner of explaining them, differ from each other.* Unfortunately one is forced, by the small value assigned to psychological observations compared to general principles, to such remote and amazing discussions, which the subject-matter itself does not at all require. Henceforth, I shall refer no more to the theory, and shall discuss the experiments only from the points of view which arise *directly out of them.*

I did not express my attitude towards the general theory of association when discussing the chance-theory, and, at the very beginning, it was pointed out that the question to be answered in this book might be affirmed or denied without thereby affirming anything regarding the relation of the experiments to the doctrine of association. For the time being, this will be assumed. If we accept the doctrine of chance, we shall also have to accept that animals have no insight whatever ; this touches the very core of the investigation. Association theorists know and recognize what one calls insight[1] in man, and contend that they can explain this by their principles just as well as the simplest association (or reproduction) by contiguity. The only thing that follows for animal behaviour is that, where it has an intelligent character, they will treat it in the same way ; but not at all that the animal lacks that which is usually called insight in man. I can, therefore, dispense with any closer elaboration in this direction and will merely observe here that the first and essential condition of a satisfactory associative explanation of intelligent behaviour would be the following achievement of the theory of association, to wit : what the grasp of a *material, inner* relation of two things to each other means (more universally : the grasp of the structure of a situation) must *strictly* be derived from the principle of association ; " relation " here meaning an *interconnexion based on the properties of these things themselves*, not a " frequent following each other " or " occurring together." This problem is the first that should be solved, because such " relations " represent the most elementary function participating in specifically intelligent behaviour, and there is no doubt at all that these relations, among other factors, continually determine the chimpanzee's behaviour[2]. They are not facts merely of the type " sensations " and the like, merely further associable elements, but it can quite definitely be proved (and quantitatively proved)[3] that they determine in a very marked degree the chimpanzee's behaviour, i.e. his inner processes, by their functional properties. Either the association theory is capable of clearly explaining the " smaller than," " farther

[1] The German word *Einsicht* is rendered by both " intelligence " and " insight," throughout this book. The lack of an adjective derived from the noun " insight," apart from other considerations, makes this procedure necessary. [Tr. Note.]

[2] As they determine memory in man too. (Cf. Selz, *l. c.*).

[3] Cf. *Abhandl. d. Preuss. Akad. d. Wiss*, 1918, Phys.-Math. Section, No. 2.

away than," "pointing straight towards," etc., according to their true meaning as mere associations from experience, and then all is well; or else the theory cannot be used as a complete explanation, because it cannot account for those factors primarily effective for the chimpanzees (as for man). In the latter case only a *participation* of the association-principle could be allowed, and at least that other class of processes, relations and *not* exterior connexions, should be recognized as an independent working principle as well.

The following explanation, which is often suggested by non-professionals, but which none who has had much experience with animals will take too seriously, can be dealt with much more shortly. Could the chimpanzees, perhaps, prior to the experiments, have seen similar methods of procedure carried out by human beings, and do they not simply imitate such proceedings ?

This idea must first of all be brought into clear relation with the question dealt with in this book. It should only be brought forward in the form of an objection if " mere imitation " means imitation *without a trace of insight* into things that have been seen ; for otherwise, instead of an objection, we should be dealing with a very special suggestion as to the interpretation of the intelligent action we have actually before us. I presume that even this slight explanation of the so-called objection will somewhat lessen the tendency to bring it forward. For any sudden introduction of relatively complex proceedings, seen *without a trace of insight*, but now performed just as if they were intelligent, would constitute a phenomenon which, as far as I am aware, has never yet been witnessed either in human or in animal psychology ; it would have to be introduced here as a new hypothesis. It appears to me, therefore, that we are faced with the following mistake. For the adult human being nothing is easier, in general, than to " imitate " what he sees, or has seen others do ; and particu-

larly such actions as the chimpanzees here carry out would
be copied immediately by one human being from another,
if occasion arose ; in such cases we may certainly speak of
" mere imitation." Now this fact, carelessly considered,
might lead to the said objection ; but when applying it to the
chimpanzee, one leaves out of account that the human imitator
has long been acquainted with the action, and, as long as the
model does not become too complicated, will immediately
*understand and intelligently grasp what the action of the other
means*, and to what extent it is a " solution " of the situation in
question. However, that it may be possible, even after a
lapse of time (for the experimenter excludes all opportunity
for imitation immediately before the experiment),[1] to achieve
complex methods of behaviour in no wise and in no detail
understood, as clear and complete actions, simply because
they were witnessed once or several times before : I repeat :
none of our experience has shown us this, and there is little
prospect that it will in the future show us anything so
remarkable. What is really important is that we consider
carefully, and allow not the smallest trace of the insight
type to be included in what we are here assuming under
" imitation."

Even animal psychologists have not always paid sufficient
attention to this fundamental difference between " simple "
human imitation and the imitation we so lightly expect from
animals, and so people were to a certain extent astonished
when it was first shown experimentally that animals do not
so easily imitate as expected. Less astonishment would
perhaps have been felt if it had been realized that, after all,
man has first to *understand*, in some degree, before it even
occurs to him to imitate. Now we have to test whether
animals also require a certain minimum of understanding of
what they have seen, before they can imitate it. Recent

[1] Excepting in those cases in which " imitation " is to be investigated.

experiments by American investigators[1] have proved quite
definitely, contrary to Thorndike's results, that some imitation,
clumsy and laborious enough, occurs among the higher
vertebrates. Their reports bear out the assumption that, in
general, the animal must work *hard* to gain some understanding
of the model, before it can imitate it. " Simple imitation " !
I can only say to any who have not yet experimented with
animals : when any animal suddenly does manage to imitate
a performance enacted before him of which he knew nothing
before, he inspires the greatest respect immediately. Unfortunately this is a very rare occurrence even among chimpanzees,[2] and when it does occur, the situation, as well as
its solution, must lie just about within the bounds set for
spontaneous solutions. It will now be seen how far removed
from experience an objection of the " simple imitation "
type is.

[Chimpanzees (and also other higher vertebrates) will
" imitate " with ease as soon as the same conditions as those
required in man are present, i.e. if they are already familiar
with, and understand the action to be imitated. If, *in such
circumstances*, there is any reason to watch the model (animal
or man), and if his actions are of interest, then either the
animal " takes part " or " tries the same solution," etc. Thus,
in imitation, similar circumstances and qualitative conditions
seem to exist in the higher animals as in man. It can easily
be shown that humans do not " simply imitate " either, if
they do not sufficiently understand an action, or a line of
thought. I shall return to this subject when describing the
imitation of chimpanzees.]

Anticipating later accounts, I will, for the present, mention

[1] Berry, *Journal of Comp. Neurol. and Psychol.*, 18, 1908 ; Haggerty,
ibid., 19, 1909.
[2] Compare Pfungst, *Bericht uber den 5 Kongress. f. exper. Psychologie*,
1912, p. 201. Pfungst, however, goes too far ; even human beings are
imitated by chimpanzees when necessity arises, *if they are understood.*

only briefly that four kinds of imitation occur in chimpanzees, but that none of the observations give the slightest ground for thinking that the animals could " simply " and *quite without insight* have " imitated " important parts of their performances. The chimpanzee cannot do this.

For the rest, the following remarks may be useful in tracing, for the present, the limits of what might be taken over in imitation of whatsoever form :—

1. The question whether the animals could ever have seen anything similar to their performances carried out by human beings is doubtless to be answered in the affirmative in some cases ; or rather, the animals *must* have seen some of these acts before their tests, though it cannot be ascertained how much attention they paid. It is, for instance, almost impossible to keep a chimpanzee in captivity without someone in his presence doing something similar to his use of a stick. Even the cleaning of his cage with brooms, and so on, must, unless some very complicated system is to be introduced just on this account, lead to similar actions. Attempts to forbid the keeper to use things in this way are useless, as first, it would be too late, because the same things may have been done on board ship, and secondly, it is very difficult for non-experts to refrain from all such actions, because men use simple implements quite unconsciously. These things have to be taken into consideration. It is less probable that they have seen boxes, and such like, used as stools, but, on the other hand, they may quite likely have seen ladders used. How far such examples influence the ultimate action of the animals, when no *immediate occasion* or incentive to imitation arises, will be discussed in another connexion. I will now merely draw attention once more to the fact that where there is no trace of understanding, the presence of the chimpanzee in

cases where implements are being used, seems to have no effect whatever.[1]

2. In a number of cases any kind of imitation is, from the nature of the case, impossible : (a) because the task in question may never have been performed by man in the presence of the chimpanzee (remember the use of a door, the unburdening of the box filled with stones, the experiment described above with the string running obliquely to the bars, and others) ; (b) because no human being would ever hit on the solution attempted by the animals (for instance, the jumping-stick and the " good errors "). Who could at any time have given them an example of how to place a box high against a wall, or to hold two sticks together to make one longer one, a purely visual solution ?

On the other hand, I must emphasize the following : It has been maintained that the chimpanzee *never* takes over a human method of procedure. That is not correct. Cases occur in which the greatest sceptic would have to admit that the chimpanzee does take over new performances, not only from his own kind, but from man also.

---

[1] I can state with absolute certainty that *no intentional instruction* of the animals ever took place, with the exception of those cases in which I personally tried my utmost to obtain some result by doing so.

VIII

THE HANDLING OF FORMS

In all intelligence tests which apply to an optically given situation, the subject of the experiment has—if one considers the problem well—among other tasks, to grasp certain forms and *shapes* ("*Gestalten*": v. Ehrenfels, Wertheimer).[1] These factors of form in most of the experiments described have been of the simplest, so that the uninitiated hardly recognize the characteristic properties of "shapes" (*Gestalten*) in them : sheer distances (very often), the relation of sizes to each other (in the experiment with the double stick, the relation of the two openings), crude directions and at the most the components of direction (model experiment of the preceding chapter, experiment with door, etc.). But always where a problem of form made greater demands on the animals—i.e. where, untheoretically, one would for the first time speak of forms and shapes (in the narrower sense) —the chimpanzee began to fail, and, regardless of fine details in the structure of the situation, to proceed as if all forms were given him "*en bloc*" only without any more precise structure. This occurred in the experiment with the wound gymnastic rope, with the coiled wire, and in building with boxes. Now, situations in which one tested mammals, from cats upwards, for intelligence, usually contained very complicated forms, especially all sorts of door-bolts and such-like. That animals below anthropoids do not immediately (if

[1] I do not quote v. Benussi in this connexion, in spite of his excellent experiments, because I find it difficult to apply his peculiar views on the question of *Gestalt* (theory of production) to the investigation of animals.

ever) understand these arrangements is obvious after what has been said. I cannot make use of such accidentally complicated experimental material when going over to more difficult experiments with the chimpanzee ; and the following tests are directed as much as possible to examining the primary functions of ever-rising degrees of difficulty, which generally remain hidden to the experimenter who makes experiments on " unlocking ", " double-bolts ", and so forth. The points to consider when planning a test are psychological, and not technical, ones ; when an animal cannot undo, or can somehow undo, a complicated fastening, the psychologist still remains entirely in the dark as to what, psychologically speaking, it was or was not able to accomplish.

The following experiences show in what direction one has to proceed so as to discover more complex situations, which will yet be sufficiently clear for the observation and the comprehension of functions :—

(March 2nd, 1914) : Tschego made her first experiments with a stick, pulling fruit with it towards the bars of her den. Now the lower part of the bars are covered with fine-meshed wire-netting, and the animal cannot get hold of the fruit which she has drawn towards her, although it lies so close, either through the tight meshes, or over the netting, which is too high for her arm to reach over it to the ground. About one metre further along, the net is lower : after Tschego has once reached down in vain, she seizes the stick again, pushes the fruit with one clear, continuous movement sideways to where the net is lower (that is, away from where she is sitting), quickly goes to the place, and seizes the fruit without further ado.

Sultan does much the same thing (March 17). The stick is tied to a rope and this is nailed to the frame of the bars. Outside, opposite him, lies the objective, but again the lower parts of the bars are covered with a fine wire-netting, so that the animal, in spite of reaching over it with his long stick, cannot touch the objective once he has pulled it straight towards him. Sultan takes the stick and pushes the fruit sideways, likewise with a determined movement ; he pushes towards a hole under the wire-netting, from where he can touch the ground outside by stretching out his arm. It is very illuminating, especially for the theory of chance, that Sultan, after he has begun to shove the fruit most carefully towards the hole, lets go the stick, goes to the hole, stretches out his arm for the fruit, and, when he still cannot reach it, immediately returns to the stick, and shoves the fruit a little closer to the hole, so that he can get hold of it from the opening.

[If the animal had not been working from that spot at the bars opposite to which the objective lay, but right from the beginning at the spot from which he afterwards reached out his hand, he would have been facing the objective sideways, and would have pulled this to him almost in a straight line, without the indirect procedure described. In order to hinder Sultan from doing this, the stick was so fixed by means of the rope that he could not use it from this second spot (that of the hole) the rope not being long enough. In the actual experiment, both animals work at an angle of 90° to 180° away from themselves, if we take 0° to be the direction objective-animal, in which, in the ordinary way, the stick procedure would have been accomplished. We thus have the case, as before in roundabout-way tests, that an act in itself meaningless, even disadvantageous, becomes intelligent in connexion with another, but only then (" Go later to the second place and reach objective from there "). In fact, the whole taken together constitutes the only possible solution. I have in a former section stated that these circumstances are characteristic of roundabout action, but I did not, at that time, wish to draw any conclusions. After the discussions in the foregoing chapter one question is at least justified.

The first part of the experiment (*a*) (" Pushing away from the animal to another place ") cannot arise intelligently *alone* ; for *alone*, it is more disadvantageous than useful ; part (*b*), however (" Going to the second place and seizing the objective "), does not yet come into consideration. Is it conceivable that (*a b*) spring from the situation intelligently surveyed by the animal (or man) as one *complete and unitary* plan of action ? I see no other way, if the beginning of the procedure, taken separately, contains no trace of a solution, but seems rather to prevent one, and so cannot arise as an isolated part. Actually a whole is required to justify, as it were, its " parts "—for such procedure as described to be intelligently accomplished. The theory of form[1] recognizes wholes which are something more than the " sum of their parts " : here a whole is required, which even stands in a certain opposition to one of its " parts ". That seems peculiar ; evidently this state of things would be crucial for any theoretical attempts to understand the occurrence of intelligent solutions physiologically.]

Functionally considered, the behaviour here observed raises two relatively simple points of view. It might be said that the animal knows how to take a roundabout way with an implement, as well as with his own body—though this possibility does not actually occur in its pure state in the test ; and, secondly, that the stick is used in reference to a later and totally different action (altering position of the body), which can only take place afterwards, as the finishing off part of the experiment. I will now deal more in detail with the *first* possibility.

It may well seem that this, where the demands on the animals should be ever-increasing, is not the place to discuss

---

[1] *Gestalttheorie* is usually rendered in this book " theory of shape " (or form), though there is no exact equivalent, and the theory is usually known by its German name. [See Tr. Preface.] (Translator's Note.)

the first point. As the very simplest form of the general type
of test, roundabout-way tests can be applied to dogs and, in a
limited degree, even to fowls. Many people may think,
therefore, that it is not of importance whether a roundabout
route is achieved with an implement in the hand, or by the
body of the animal ; if, in the former case, the animal is
familiar with the use of the implement, the making of detours
—well-known from its own movements—should almost be
self-evident. In fact, this might follow in any intellectualistic
conception of the nature of intelligent conduct. But here
the same thing happens as otherwise in higher psychology :
even intelligent behaviour, the achievement of insight, will
not submit to " intellectualistic interpretation." At any
rate, the chimpanzee is very far from using the roundabout
methods with implements (any objects), as easily as with his
own body.

I shall describe tests in this direction which were first
performed on the quietest, most carefully proceeding animal,
Nueva. She sits behind a railing, outside which, forty-five
centimetres away, is a contrivance in the form of a square
drawer (open on top), from which one side is missing. The
edges are thirty-eight centimetres long, the three vertical
sides six centimetres high ; this " roundabout-way-board "
is placed on otherwise free ground, in such a way that the side
without a vertical wall is turned away from the animal (herein-
after called the normal position) (cf. Fig. 17). The experimenter places the objective (banana) at point O, and then gives
Nueva a rather long stick (March 18th). The animal scratches
the objective towards her (o°), but soon cannot get it any
farther because the front side of the drawer is in the way.
She becomes very distressed, complains, and pleads, but
no help is forthcoming. At last she seizes the stick again and
tries once more to pull the objective towards her at o° (i.e.
in a straight line). Suddenly she changes her tactics ; *instead*

of putting the stick *behind* the objective and pulling, she puts it *in front* and *pushes* it with little jabs, but with all assurance towards the open side (that is, in the direction of about 180°). She keeps up this careful and regular shoving until near the edge of the board, where, without any jerk or unsteadiness in the conduct of the animal, the stick happens to be brought *behind* the objective, which is pulled back several centimetres (about five). The " change " only lasts a few moments and

APE

FIG. 17.

then she starts pushing quite obviously towards the opening again ; the objective is quietly pushed along sideways from the drawer with even movements and finally brought to port in a curve (on the left side from the ape).

On repeating this performance a few minutes later, the whole detour, with a clear beginning at 180°, is again accomplished without any mistake.

At the following day's repetition, Nueva begins by drawing the objective closer at 0°, then, quite suddenly, she changes her direction before the obstructing side is actually reached, thus smoothly pushing the objective over a large portion of the board *away from herself.* She alters the direction for a moment, as on the preceding day, and, after that, carefully and smoothly describes the curve of the detour. (After a few minutes it is again done, easily and without mistake.)

(20.3) The board has an area of fifty centimetres square, and so the circuitous route to be taken is longer. Nueva starts at o°, and again, before reaching the side, changes suddenly round, and pushes the objective calmly and with care by the circuitous curve to within reach. (Repetition after a few minutes : correct solution.)

[On March 28th, the test is made once more. Nueva begins at o° and abruptly changes over to 180°. When, in bringing the objective round the corner of the board, its side gets into the way of the stick, the animal quietly, but decidedly, pushes the whole board aside with the stick, and continues her work comfortably.]

Nueva's behaviour in this test is much clearer than anything which will be reported later of the other animals, and yet it shows clearly enough that the only kind of solution which comes into consideration here, and which is actually achieved (after more primitive behaviour at the beginning), can only succeed against some strong resistance. There is no doubt that Nueva's solution is an intelligent one : the new direction (180°) is distinguished clearly from the first one (o°) and there is no scattered trying-around at all. But that it takes such a long time to discover the solution, and that the animal, after the first primitive effort, remains perplexed for a time, that, even after six attempts, the direction (o°) returns first before the movement in the right direction is begun—all this is in sharp contrast to the matter-of-fact way in which chimpanzees *run* or *climb* roundabout ways to their objectives. The curious " change " which is still observed even at the third experiment (on the second day) further shows that it remains hard to accomplish this solution, even after it has once appeared clearly and has been carried out quite a distance. This momentary and spatially very limited backward movement has no element of uncertain trying-around in it. I can best characterize its nature by an approximate analogy :

if a man has to execute movements (which ordinarily he can perform with ease) when observing them in a mirror, it often happens[1] that he is brought suddenly to alter the directions of his movements, as by some force, because the normal adjustment between the visual and motor factors is disturbed. When Nueva reverts, at times, to the normal direction, pulling, the observer gets the impression that the animal itself is only made aware of the change after she has covered a part of the way at (o°). In later experiments, this phenomenon not only occurs again, but is exaggerated to such an extent that it almost becomes paradoxical.

Only one other animal, the clever Sultan, achieved a solution at all, when the board was in its *normal position*. How he managed it is remarkable not only on account of the unpleasant difference as compared with Nueva's experiments (18.3). The board thirty-eight centimetres square is used, and is placed a little farther away from the bars (fifty-five centimetres). Sultan pulls the banana towards him (o°) and endeavours to lift it over the edge ; but, as the side of the drawer makes it altogether unattainable with the tip of the stick, the observer puts it back in its original place. Sultan now moves it sideways (about 90°) to the wall, and when the objective has reached this, begins to lift it with the tip of the stick, and really pushes it out so that it is easy to pull along on the ground. The small vertical detour (six centimetres) over the edge seems to come without difficulty ; as soon as the fruit reaches the wall, *lifting movements* plainly take the place of pushing ones.

Up to now, the action has never been directed to the open side. This occurrence is now provoked by an accident which proves in general of strong assistance. A new objective is provided. With the hasty movements which in this experiment distinguish Sultan unfavourably from Nueva,

[1] If I am not mistaken, this experiment was originated by Mach.

and become more and more unordered after many useless efforts, the elastic banana jumps from the board a little way and, in falling down, rolls away in the direction of the open side. Sultan immediately changes his procedure, *pushes* the objective farther *out* obliquely, and then draws it to him in a curve. Exactly the same thing happens at the next repetition ; at first the animal works as if quite ignorant, in directions between 0° and 90°, until suddenly, under the strong pressure of the stick the banana bounces away from him towards the open side ; again in the same moment Sultan changes his tactics and solves the problem. Of course it had become easier, since the banana thus, accidentally, approached the open edge, and the curve did not need to start from the direction 180°, which in the experiments on the other animals proved to be a particularly difficult one (compare below).

(19.3) In order to make any chance assistance more difficult, the small board is replaced by the bigger one, fifty centimetres square, but the procedure remains the same : Sultan tries to lift the objective sideways over the edge ; it bounces away several times, and when it finally rolls close to the open side, he changes over abruptly to the correct movement, and gets the objective into his possession without further trouble. The next time, nevertheless, he again starts by his first method —pushing towards the side ; this time the banana does not bounce so near the open side, but back to the middle of the board ; this movement seems to act suggestively, for suddenly Sultan works away at 180°, and achieves a perfect solution. The third experiment of the day required no further assistance from chance, the objective is straightway shoved off the board without any mistake, and drawn round in a curve.

After an interval of two months (16.5), the animal first starts off in its original direction (0°) ; then stopping sharply, the correct solution appears in a faultless curve.

After the final form of the solution, and after the way that the assistance of chance is each time used, I must consider the performance in its final state as intelligent behaviour, even if it is striking that chance assists the animal three times to a perfect solution, and he is yet not able to produce it on its own next time, or even to indicate it. This only seems possible if some strong force is working against the solution, or, to put it more exactly, if such a force is preventing the beginning of the solution (direction 180°) from occurring to him. The second expression is more appropriate, because only the *beginning* in the difficult direction has to be assisted by chance, for Sultan to jump to the *whole* solution. (The last follows directly from the fact that the curve is described as " roundly " as possible, *every time* ; while still on the board the objective near the opening gets that sideways-slanting component of movement which corresponds to the further continuation of the curve on free ground, i.e. the " moving-around.") As to the nature of the accidental aid, several interpretations are possible, the experimental tests of which have still to be made. The solution is brought about either by *the proximity of the objective to the open side*, after it has bounced, or else the deciding factor is the *dynamics* of this jump in the difficult *direction* of the commencement of the curve, or, finally, it is the effect of both together. I consider the last correct ; but according to all other experiences with animals and human beings, the most probable is that *the movement itself with its inherent direction-factor, constitutes the chief suggestive force.*

[It may further be asked how the complete solution-curve can be produced in this manner. To this again, two answers are possible : either one can imagine a connexion by association which, already existing in the animal, by the reproductive force of the chance jumping away, can call into being the whole curve—or there exist, so to speak, " autochthonous " possibilities in the animal, which effect the sudden appearance of the solution-curve in the new total situation " suggestion of direction in the given structure " ; the arising of this solution, in the original static situation, is only prevented by strong opposing forces. This second assumption would in all cases of clear solutions *without assistance* (for example, the conduct of Nueva in a similar test) include the hypothesis that the directions, curves, etc., of these solutions, could arise autonomously (not necessarily " from experience ") in the static situation. According to the plan which we have proposed to ourselves in this book, I leave the choice open.]

The numerous tests with other animals need not be reported in such detail, as they only differ from the ones described, inasmuch as the difficulties of the task come out still more definitely. This fact will appear more clearly in a briefer survey :

Chica.

(18.3 and 20.3) Normal position of the board.	The direction 0° is steadily adhered to.
(18.3) The board is turned as shown in Fig. 18.	Chica is so violent, that the objective bounces and jumps towards the opening ; immediately thereupon the solution occurs. (At a repetition, the solution occurs only after the same assistance of chance.)
(20.3) Same position.	Direction 0° to begin with ; banana jumps away, solution follows. In two repetitions, clear solution from the very beginning. (However, compare below.)

Two months later (16.4) Normal position.

Direction 0°, objective actually lifted over the edge.

At repetition, direction 0° is maintained, in spite of strong chance aids ; almost from the open side itself, objective is brought back at 0°. But suddenly, sharply distinguished, the solution occurs (180°, and so forth).

In two further repetitions the circuitous curve is entered upon correctly from the beginning ; but at the same time there are several " sudden reversals ", such as we are used to in Nueva (by no means mere " trying-around "). Last repetition : even this disturbance disappears.

In the experiments of March 20th, Chica obtains for herself very characteristic aid. She does not, as she used to, work from the ground, but sits down on a cross-beam of the bars about seventy centimetres up, not in the middle of the arrangement, but at point C (cf. Fig. 18). One can see at a glance how the circuitous route is thus facilitated, and not only from a motor standpoint.

Grande.

(18.3 and 14.5) Normal position.

Direction 0° is persisted in, in spite of chance aids. Grande beats the board with rage.

(14.5) Quarter turn to the left.

Grande keeps the primitive direction.

Further quarter turn. (Opening to the side.)	Problem solved at once at 90°.
Quarter turn backwards.	This problem, too, perfectly plainly solved now. (Direction 135°.)
Normal position.	From the beginning 180°, and faultless solution.
One month later (18.6) Normal position.	Clear solution from the very first moment.

o o o o o o o o o C o o

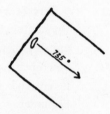

FIG. 18.

[Grande tries at times to shorten the proceedings by drawing the whole board with the stick, or with her free hand, towards the bars. The direction 90° occurs for the first time with this animal when the board is lying at right angles, turned sideways; it occurred immediately under these conditions. That the solution is afterwards quite naturally transferred to the two more difficult positions, although they require modified movements, that this change is taken into consideration, shows the grasping of " structural relation." A further test was added to the one on May 14th (when the board was in its normal position), by giving the board a quarter turn *to the right* from its normal position; the solution came at once, the curve being made *to the right* corresponding to the change of position (right and left being here always reckoned from the animal).]

Tercera.

(18.3 ; 20.3 ; 18.6) Normal position.	Direction always 0°, although there are chance aids.
(20.3 ; 18.6) Quarter turn to left.	Clumsy movements at 0°, in spite of chance aids ; the animal looks extremely stupid and lazy[1].
(18.6) Further quarter turn to left (opening to the side).	Solution occurs immediately at 90°.
(18.6) Quarter turn back again.	Tercera begins at 0°, discovers the solution immediately by chance aid (beginning at about 135°). In two repetitions the circuitous curve is entered upon from the very beginning.

Tercera, who is usually very lively, but immediately falls into a kind of stupor when she is to perform experiments, shows quite a striking difference between the stick movements *before* the beginning of the solution and *after* the critical moment (e.g. after the aid of chance) ; at first, she vaguely fumbled, but her movements became precise the moment the direction of the solution appeared. Though her working remains always clumsy.

Tschego.

(20.3) Normal position.	Direction 0° without any deviation.

---

[1] Contrary to Rana, who tends to be stupid and active.

Quarter turn to the left.	The direction remains for a long time 0°, until Tschego finally in a great rage breaks the stick to pieces on the board.
Further quarter turn to left (opening at side).	Tschego remains at 0° for a while, then suddenly changes over to a clear and careful solution (i.e. beginning with 90°).
	On repetition the direction is again 0° to begin with, but changes abruptly to the right one.

During Tschego's solutions, a remarkable motor phenomenon takes place ; when the objective is already nearly at the opening, the animal changes the stick from its right to its left hand, presumably because the right is tired, and now for a moment performs with its left hand movements in symmetrical relation to the foregoing ones (i.e. to the right at 90°), so that the banana is pushed a few centimetres back into the drawer. This error, it is true, is immediately corrected, but occurs again every time the stick is changed from left to right hand for a moment. This phenomenon has nothing in common with the sudden change from the new direction to the biologically primitive one, noticed in Chica and Nueva, but may be due to the co-ordination of motor functions of both arms, which with us too, will often make *symmetrical* transfers from one side of the body to the other in preference to *identical* ones.

Rana

(19.6 Normal position.	Works all the time at 0°.
Quarter turn to left.	Remains at 0° without any deviation.

| Further quarter turn to left (opening sideways). | Rana keeps on for a little time at 0°, but afterwards goes over to the solution. On a first repetition the same procedure takes place, i.e. beginning at 0° and later transition to 90°; on the second repetition Rana keeps obstinately to the primitive direction and does not leave it, even when the objective is quite close to the opening. |

These results prove clearly enough that the performance required here is incomparably more difficult than ordinary roundabout ways. If we were to bring any of the chimpanzees into a square room, entirely railed off except for one wall, but in the same proportion of size to the body of the chimpanzee as the roundabout-way board is to the banana, and if the animal were to stand at the spot opposite O (cf. Fig. 4, p. 15), it would perhaps try for a moment to reach between the bars, but would certainly soon start off determinedly on the detour whose beginning is at 180°. The solution would, therefore, come about in a " normal position ", without our being obliged to facilitate the problem for any of the animals (as we have just had to do for the majority) by quarter or half turns of the cage. Even a bright dog, as we have seen (compare above, p. 13), can easily achieve the same in any unknown roundabout situation set up *ad hoc*. The fundamental difference that appears here can, in spite of the simplicity of the experiment, be explained by different factors : first of all, the making of the detours (compare above) may be so much more difficult with a tool than with the animal's own body ; but the difference might also have to do with the fact that the *detour must be made, not from the standpoint of the animal towards the objective, but the other way round, from the original place of the objective towards the animal.* In order to decide the theoretically-important question, which factor is the salient one—for both probably act together—detour experiments should be made where the implement (stick) has to be used from the animal *towards an objective.*

The added facility given by turning the board sideways is quite plain; even at 135°, the detour curve is more easily entered upon (Chica), and when the required movement has to start at about 90°, all the animals sooner or later suddenly come upon the solution. We shall have to consider carefully what interpretation to give to this dependence on the " geometry of the situation " (compare also pp. 15 and 36 seqq.). In this matter the detours just described are identical with the ordinary ones made with the body; one has only to test hens instead of chimpanzees for it to be shown that, for them, the detours that begin at 180° away from the objective are altogether impossible as genuine solutions; and that it is more likely that the task will be accomplished on approaching 90°.[1] To the human observer it is obvious from the beginning that the board test must succeed somewhat more easily when beginning at 135°, and much more easily at 90° than at 180°; and, this time, experience supports him. It is not so easy to say wherein the difference lies; perhaps the detour curves will strike him as different in smoothness. But what does this mean psychologically, and in how far does it determine the different degrees of difficulty ?

The most striking phenomenon in these tests is still the sudden occurrence of perfectly clear and definite solutions, when once a *single* chance movement has brought the objective a little in the direction of the beginning of the curve; it is as if, at least temporarily and for that experiment, a spell

---

[1] Even at 90° the obstacle must require a short detour only, if hens shall be able to see the solution-curve.

had been broken. Only the more foolish animals can never be helped even in that way.

[I imagine that no one will wish to play off the frequent occurrence of favourable accidents in these experiments against the considerations of the previous chapter. This is actually the first case in all the observations in which they occur, and it is seen easily enough that the physical movement, which from the standpoint of the animal has to be considered accidental, *must* occur frequently here (whilst in other experiments such one-sided favourable conditions do not exist). The fruit bounces away, in the first place, when the animal is trying to lift it over the edge ; if, during this operation, it falls, as it generally does, from the narrow stick, obviously the direction of the fall is away from the animal, because the stick runs slanting downwards from the animal's hand. Secondly, the fruit bounces away, when the animal, instead of putting the stick on the ground behind the objective, puts it hastily only on top of the fruit, pressing it a little, and then pulls ; the board (in contrast to the ground) is smooth, and if the pressure is at all clumsy or, in excitement, too strong, then the stick will slip off frontwards, and the fruit must bounce away.]

He who reads the description of the experiments attentively, will realize that the performances of the different animals decrease in merit in the order chosen here. (Grande is distinctly better than Tercera, because of the ease with which she produces the solution on turning the board back.) The animals classify themselves in this same order : Nueva, Sultan, Chica, Grande, Tercera, Tschego, Rana, apart from these special experiments too. i.e. if one is determining the degree of their intelligence according to their whole behaviour and the character of their other performances. I only noticed while writing this section, that this board test gives the animals the same places I had already attributed to them long before in my mind. (Tercera I put between Grande and Tschego with some uncertainty before, as she was so seldom to be induced to serious effort in experiments ; but the board test justifies me.)

[Koko is not included in this classification, as the weakness of his arms hindered him very much in directing the stick in the board test, and the uncertain movements were harder to judge. But he undoubtedly first worked at 0°, like all the others ; with him too the objective once bounced towards the opening, and he then tried, but with no real success, to push it farther in the direction leading to the solution. According to this, he ought to be put equal to Sultan, whose level of intelligence and character he was also nearest to in other ways. Konsul was not tested.]

As far as method is concerned, it follows that in some cases the intelligent treatment of optically given situations can be tested by methods which have a certain resemblance to the working methods of the psychology of perception. (Visual perception of forms in space, of movement, etc.) This work contains only the first beginnings of these experiments, as the animals only gradually drew attention to such possibilities by their behaviour.[1]

[For comparison, I will relate an experiment in which a boy of two years and one month was tested exactly like the chimpanzees. The child may be described as of average intelligence. He is put into a railed-off space, such as is often used for little children. The walls are so low that they reach only to his breast. Inside lies a light stick ; outside, out of his reach, the objective. In a little while the stick is picked up as a matter of course, and the objective pulled to him

---

[1] In future it will be better to use only the vertical sides of the round-about-way board, and *not its wooden* bottom ; perhaps the animals can make the roundabout way more easily on free ground than over *wood and ground ;* the sharp contour of the wooden board as against the ground may also make the test more difficult.

with it. The skill with which this is done is distinctly less than that of Sultan, who is twice as old as the child, but greater than that of Rana and Tercera, who are about the same age as Sultan. In whatever way the use of the implement may have developed, it certainly takes place.

On the same day the board experiment is made and in its normal position. The child again immediately takes up the stick, but proceeds so clumsily that he drops it before he has used it. He puts his foot through the bars on to the stick outside and pulls it closer, but does not bring it inside, perhaps because he does not see how a stick lying crosswise can be got through the bars. Instead of that, he hits at the stick with his belt, which has fallen, then stands for a while looking dejected, and makes the onlooker understand that he wants the stick. This is handed to him. The boy takes it, pulls the objective straight towards him with it at 0°, keeps on thus for some time, though the objective bumps repeatedly against the side wall of the board, and finally changes over to the left-hand corner from himself (about 45°)—the observer meanwhile having put the objective in its old place. After many useless efforts, the child gives up the work. He takes the stick and throws it at the objective, then the belt also flies outside ; if he had had anything else to throw it would certainly have gone the same way—just as in the case of the chimpanzee (compare above, p. 88 seqq.). It was proved directly afterwards that the child took circuitous routes himself (i.e. with his own body) without trouble ; a much younger child had, moreover, been tested with success in this respect before (compare above, p. 14).]

The task where the animals have to deviate from the direct path, in dealing with objects, and, instead, to adapt their direction of procedure to the forms before them, can be examined in many experiments differing from each other externally. I will give one more example, in which the forms which have to be taken into account are somewhat different.

In the introduction an experiment was described in which the animal had only to draw a ring (or a loop) from the stump of a bough (or nail) in order to make the objective fall to the ground, where it could be easily picked up. Actually the ring (or loop) was not noted at all, perhaps because the connexion between the way the ring was fixed and the rest of the situation was not grasped ; the animal did not get as far as taking an interest in that. A situation is now prepared in which the animal, as far as one can see, must make an immediate effort in order to find a solution to a connexion of this kind.

On the other side of the bars lies the objective, out of reach. A stout cord is fastened to a stick with which the animal could reach the objective ; on the free end of the cord is a metal ring of about six cms. diameter, which is slipped over a nail sticking out vertically about ten centimetres from a heavy case. With the string stretched, the stick does not reach even to the bars, and, therefore, in order to be used, the ring has to be taken off the nail with a movement deviating by 90° from the primitive direction " stick direct towards bars and goal ". This movement can be " genuinely " accomplished only if the animals are able to grasp the arrangement " ring over nail ". Those who have not seen how chimpanzees deal with more complex forms, may think that there could be no easier task.

(21.2.1914) Sultan pulls at the stick in the direction of the bars (and of the objective), chews and gnaws at the rope where it is tied to the stick, notices the connexion ring-nail only after a considerable time, and then does not lift the wide open ring up a few centimetres, but tries to pull out or break off the nail ! *The final solution is that the stick itself is broken above the middle with a great effort, and the objective reached with the free part.*

On repeating the test with a new stick Sultan notices the movements which the ring makes on the nail (when pulled towards the bars); he touches the ring as if examining it, and then takes it off with one quiet clear movement. The next time, none the less, he pulls first of all towards the bars before turning to the ring and pulling it off again in one sure movement.

Grande, Chica, Rana, and Tercera first pull at the stick and endeavour persistently to solve the connexion " rope-stick "; under their impatient movements the ring on the nail gets out of position, and it even slips off; but the animals do not notice this in their absorption in disconnecting the stick from the rope, and the ring could always be put back on the nail when they were not looking. The following is the limit of this behaviour : Rana accidentally pulls the ring from the nail, sits close to the bars, not noticing that now the stick is free to be used; the observer again puts the ring back over the nail unobserved by the animal, and immediately afterwards Rana pulls in the direction of the bars. When the same accident again occurs, and the rope hangs in the air with the ring free, the animal realizes only after a while that the stick can now be moved freely and that the connexion between it and the *rope* does not matter any more. The animals named did not at that time reach a genuine solution of this problem.

As the chimpanzees want only to have the stick and, as the next part of the whole arrangement, the rope, is so thin and flexible as to invite being torn or chewed, the attention of the animals becomes fixed on it to a surprising extent; efforts made to help them out of this were of no avail. Therefore, in later experiments, the rope connexion was omitted, the ring being nailed on the end of the stick, but in such a way that the greater part of the opening stuck up above the wood of the stick; and, in order to make chance solutions still more difficult of occurrence, I replaced the nail of the former experiments by an iron bar, standing out about thirty-five centimetres vertically from a heavy box.

(10.5) Rana pulls at the stick in the direction of the objective and does not take any notice as yet of the ring ; as the stick will not come off, she finally upsets the whole box by her tremendous straining in the direction of the bars ; thus the stick falls off. The observer has the impression that instead of the ring round the iron bar, any other objects of equal total size could be used ; this would not matter very much to Rana ; it does not occur to her to look at the thing.

(14.5) Rana this time pulls so hard in the direction of the objective that the iron bar in the box bends a little, and the ring slips off ; the animal scarcely understands why the stick suddenly becomes loose in her hand.—The next time when Rana pulls, the bar does not give way ; she thereupon has a good look at the critical spot, pushes the ring up a bit higher, *but directly afterwards begins to pull as before in a horizontal direction.* This clumsy proceeding goes on so long and so violently that the nails which hold the ring to the stick become bent, and the stick is let loose ! (If one behaves foolishly in such a situation, one has to pay for it with many foot-pounds of work ; the nails used in this instance were very strong. On the other hand to lift off the ring would have meant a minimum of work, and we see from this small example of what fundamental importance for a *technical* consideration of the organism is the degree in which the handling of things is determined by a clearly-grasped structure of forms in space. Entirely apart from all psychology, it is of the greatest interest to every technologist to see cleared up the properties and processes of an organism—a material system after all—which cause such far-reaching physical differences.) In the following experiment Rana, surprisingly enough, does not pull at the stick at all, but raises the ring, without further ado, over the top of the iron bar, so that one might really think that it was an intelligent action ; the experiment is at once repeated and Rana this time pulls sideways quite primitively. In two further cases horizontal pulling at the beginning is each time followed by quick and sure lifting of the ring.

(11.5) Grande is tested in the same situation. She pulls at the stick in the direction of the objective, without casting a glance at the place where it is fastened, and then, for a time, stops bothering about the task. When the other animals are fed outside, she starts pulling again, but just at this moment (no doubt by chance) she looks at the ring, and a slight upward movement of it (perhaps five centimetres) does not escape her. This works upon her immediately, like the chance assistance in the board test : Grande goes up to the thing and, with one single movement upwards, lifts both the ring and stick.

(12.5) In two experiments, one after the other, Chica at once accomplishes the solution.

It might very well be thought that, after this, the animals would in future retain this simple proceeding as an assured possession, and if the ring, which is pulled over an iron rod (nail) were a visual fact, as simple and crude as " a-box-in-the-neighbourhood - of - a - vertical - distance - which - has - to - be-bridged," the animals would really be able easily to accomplish its release. But that is by no means always the case. Sultan tries to solve (19.5) such a combination (ring-nail) but moves round about it in an aimless way and, finally with a violent movement which pays no heed to the nature of the fastening, only succeeds through sheer strength, and pulls the ring down. In further tests I have seen the same animal take off the ring (or loops of rope) with all possible care, from nails, rods, boughs of trees, but I have just as often seen him blundering around the same combinations. On one occasion later, Grande actually succeeded with great effort in accomplishing a solution by dragging the iron bar on which the ring hung out of its fastenings, rather than by the already known and apparently so simple method of taking off the ring ; the iron bar is then used instead of the wooden stick ! But, on another occasion, when she begins to loosen the iron bar, this is obviously done only in order to free the wooden stick ; and yet the ring had already slipped so far up the iron bar and somehow remained there, that a very slight lift would have accomplished the solution (19.5).

The question discussed here would not be better answered if one tried to bring about in further experiments the clear accomplishment of this small performance. By such practice, one would probably obtain the regularity desired ; but that the apes treat one and the same problem, sometimes blindly, and at the other times with perfect clearness, is just the characteristic thing about these animals. The most obvious explanation of their behaviour would seem to be that they always find the clear solution, when they clearly grasp the structure of the connexion, and, on the contrary, that they pull at it crudely when they are not able to achieve this clearness. The ring over the bar (the nail) seems to represent to the chimpanzee an optical complex which may still be mastered completely, if the conditions are for the moment favourable, if there is concentration of attention and so on ; but it has a strong tendency to be seen less clearly if the animal fails to make the proper effort on its part. We therefore here approach more difficult structures as, e.g., the " coiled rope ", " relation of shapes of boxes ", and so on, which rarely suggest definitely to the animal what movements it has to perform. Anyone who undertakes such examination will observe only too quickly that the animals do not always enter into the experiments in an equally calm and attentive way. The smallness of the spatial forms that are here under consideration might very likely add to the difficulties of the task of grasping them clearly ; the experiments that have been described up till now have usually been made in situations, whose parts are purposely interrelated in forms that are not only simple but also big.

[As presumably the structural complex dealt with often remains obscure, there cannot quickly arise a state of mechanization in which by a mere glance towards the complex the appropriate curve of movement appears. This would only be possible if the structure " ring-nail " itself could be established by practice once for all, thus creating the conditions for the reproduction of a mechanical proceeding. This establishing should be possible with the chimpanzee, according to my experiences ; but it is of no interest here.]

The observations recorded show also that we have now left the field in which the experiments give simple and decisive answers to our questions. It lies not in the experimentation[1], but in the nature of the animals, if the results become gradually less and less clear ; in the animals' visual and other brain centres also, things may well become less and less clear, as the experimental conditions reach a certain degree of complexity. If we had not made the acquaintance of the chimpanzees in optically simpler situations, we should have found it difficult now to take any position at all as to their behaviour. And yet many experiments with mammals have begun with the treatment of just such complicated situations, as if they were simple ; results in such cases must either be equivocal, or, with the increasing complications, must turn out negative, and then no conclusions of any value to the fundamental question of insight can be reached.

Variation of the experiment. At the height of a man, a rod two metres long is fixed to the wall of a house, so that it stands out at a right angle ; a little basket, with the objective in it, is hung by its semicircular handle on the rod, about one metre twenty centimetres from its free end. A little to the side, on

---

[1] At any rate I am sure that anyone who approaches similar tasks can now avoid mistakes which I made.

the ground, lies a long stick. (11.8) Sultan is brought in ;
he looks up at the basket, wants to climb up along the beams
of the house, but is hindered, and, glancing round about him,
remains squatting on the floor near by. It is only after some
seconds, when his eyes have been directed to the stick close
to him, that he seizes it and rushes towards the basket.
Twice he beats at it, blindly and simply, in the oblique
direction that his position happens to determine, then he
suddenly changes the direction by 90° towards the correct
side and, in a cautious movement, moving it six times carefully, pushes the basket towards the free end, until it falls
down.

Grande in the same situation drags a box from far away,
places it under the basket, mounts on it, but does not reach
the basket. She fetches the stick, but lets it drop immediately
for no obvious reason, and runs to a second box at a distance
of about fifteen metres. Whilst she is busy pulling it over the
intervening distance and not looking at us, the first box is
removed and hidden. Immediately afterwards, the animal
arrives with the second box, places it in position, mounts on it,
and still does not reach the basket ; she looks around with an
expression of astonishment, and finally turns to the observer
wailing. Left without assistance she again seizes the stick and,
from the very beginning, pushes the basket down over the free
end correctly and without making a false movement on the
way. On repeating the experiment, on the other hand,
Grande works the basket a few centimetres the wrong way,
towards the house-wall ; then, abruptly, reverses the movement by 180°, and pushes the basket along the stick steadily,
until it falls.

[The complaining in the midst of the experiment was not
provoked only by the fact that the animal did not succeed ;
for the looking around that preceded it was undoubtedly
full of astonishment, and the wailing carried a note of indig-

nation. The other box was missed, as soon as the need for a second building-block was felt.]

The experiments mentioned at the beginning of this section convey, besides the roundabout way of dealing with objects, yet another principle : the objective is put by the use of tools into a position in which it can be reached only by changing the position of the animal's own body afterwards. In the case described above, however, this procedure has been facilitated very much for the animals, inasmuch as afterwards they only need make one or two steps sideways, thereby remaining at the same bars at which they have worked from the beginning with the stick ; these bars, furthermore, are so well known, that " near the bars " (at whatever place) and " attainable ", " accessible-to-me ", ought to be very closely connected to the animal. One may sharpen the conditions of the experiment essentially by requiring the animal to " take into consideration ", during the use of the tools, a greater subsequent changing of the position of his own body, so that he works at first in one certain spatial orientation, for quite a different later one. The total " curve of behaviour " is formed in such a case by two lines running in *opposite* directions, whilst the experiments last considered (for instance, with the detour-board) move on the same total curve in a single movement of one direction only.

A big wooden animal cage is closed on one side by bars, between which the animals can pass their hands from outside ; but the cage is so big that the arm of a young chimpanzee standing outside does not command the whole interior from these bars, but only about half of it. The side opposite the bars consists of boards nailed across horizontally ; one board is removed, at a place that enables the young animals to look and put their hand into the cage, but not touch the floor ; the rest of the cage is closed. If a piece of fruit lies on the floor close to the wall from which a board has been removed, the chimpanzee will reach after it from the (opposite) bars with a stick, as the cage (weighted with stones) cannot be turned over. If one takes care that the stick can be used only from the side of the gap (where one board is removed), the sole solution remaining is to push the objective from the gap towards the bars, until it can be reached there with the hand, One therefore removes all possible sticks and staves except

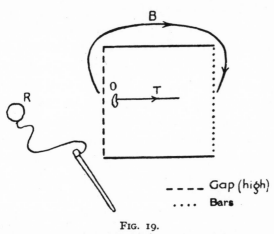

FIG. 19.

one which can be used quite comfortably from the gap, but, being fastened neaf the gap by a rope tied to a tree, cannot be taken over to the side where the bars are. (Fig. 19 shows only the ground-plan : R is the tree with the rope and the stick attached to it ; the broken line indicates the side with the gap ; opposite, the bars are indicated by a dotted line. The lines T and B indicate the two parts of the total procedure running towards each other, one of which is to be covered by the tool, the other afterwards with the animal's body. It is clear that the ape has to work for a later position of its body, which is, as it were, the reverse of the position taken during the use of the tool.)

(27.3.1914) Sultan seizes the stick, pokes with it through

the gap, and tries to pull the objective towards him and to lift it up the side to a height where it can be reached. From time to time he runs off, looking for a blade of straw, or something like it, with which to reach for the objective from the side of the bars—but in vain. After a while—the animal is again using the stick from the gap—the whole direction of the movement suddenly changes; the objective is pushed away from the gap, not to the bars, but to a spot where in one of the sides below, about half-way between the gap and the bars, there is a little hole in the wood. Sultan proceeds very carefully, brings the objective in front of the small hole with the stick, then drops the tool, goes round to the place outside the hole, and makes a great effort to squeeze out the fruit with his fingers—but the hole is too small. He soon again approaches the gap, again seizes the stick, and now changes the position of the objective in a way which I could not clearly understand, but probably still counting on that hole, and in any case getting close to it. In doing this the goal comes across the middle of the floor of the cage, a little closer to the side of the bars. All of a sudden, Sultan drops the stick, runs round to the bars, puts his arm through them as far as he can, and actually reaches the objective. The impression upon the observer after this proceeding is *not* that Sultan has immediately before worked in the direction of the bars and now comes round to complete the success thereby rendered possible ; it looks rather as if he once again had abandoned the use of the stick in order to try his luck from the bars, as he had done several times before. As the attempted solution with the hole in the wall after all contains the method required, although in a simpler form, and since a human being would probably consider the accidental success just described as a strong help to the animal, all depends now on what he does upon repeating the experiment.

A new objective is put in the place where the first one

was. Sultan seizes the stick and *pushes the fruit straight towards the bars,* without taking any further notice of the hole in the side.   On the way, one notices several times indications of the " changes " into the (biologically evidently very strong) direction of 0°[1], observed in Chica and Nueva, but never before in Sultan : inasmuch as the stick is placed, erroneously, *behind* the objective, for a moment the movement of pulling is made, and if the correction were not made immediately, the objective would return to Sultan.   As a matter of fact, the total of the small backward pulls towards himself amounts only to a few centimetres, as the animal itself soon realizes what it is doing.   Sultan makes the whole course unnecessarily long, as he does not take into account the length of his arm ; with the greatest effort he pushes the objective right up to the bars, that is to say, a distance of about one metre, and finally gives the fruit a push with the stick (which is a little too short for the operation) so that it falls out on the ground between the bars.   But in that same moment he is already running round the cage, and gets the objective.   The very deviation from his behaviour during the previous chance success (when he reached far into the cage) proves that after this bit of help a genuine solution of the problem has arisen.

At a repetition of the experiment Sultan nevertheless reaches for the objective from the side of the bars, with a straw, before he goes to the gap and solves his task.   He achieves it without a " change," but again disregards the length of his arm, and makes unnecessary efforts to push the objective with his short stick right over to the other side. During the third experiment the procedure is perfectly clear ; Sultan stops pushing when the fruit is still a good distance from the bars, drops the stick and runs round.

---

[1] It seems to me quite impossible that this is only a question of a sudden change into an " accustomed " way of using the stick.   At 90° the animals *push* the objective without any inhibitions,

Chica likewise hits upon the solution (30.3) with chance assistance. She first of all pulls the objective towards her at 0°, and when she tries to lift it up the wall, it drops, and bounces awayf rom her to about the centre of the cage-floor. At the same moment the animal runs round the cage, passes her arm through the bars, and reaches the objective.

As with Sultan, the consequence of what has thus happened is, that, in the next experiment, *from the very beginning, the direction towards the bars is clearly taken.* There is no doubt that this is the beginning of the solution. But now occurs one of the strangest performances that I have ever noticed in these animals. During the board experiment Chica had already frequently turned off from the right track (180°) and, for moments, reverted to the primitive direction (0°). While working now perfectly correctly and clearly towards the bars opposite, she is startled by a noise from the street, looks for an instant towards the scene of the disturbance, and then continues her activity, *but now pulling at* 0° ; this time, the change is *not* corrected. Chica continues pulling until the objective is brought close up to her under the wall with the gap ; and *at this moment, like some one who has nothing more to do than to reap the fruits of his efforts, she runs round the cage to the bars* ; nobody could look more nonplussed than Chica, when she peered into the cage and saw the objective as far as it could be from the bars. It looked as if she had just been awakened out of a dream, and, judging by the usual behaviour of a chimpanzee, the only explanation that can be given for this performance is that the disturbance had a long after-effect, and that, under these conditions, particularly favourable for the change, it was not noticed and corrected as usual. In this way Chica brought the objective up to the natural end of its track, and then proceeded, still " absent-mindedly ", to the second part of the programme, which now, of course, did not fit in, and thus led to the " awakening."

After the surprise, the animal returns to the wall, seizes the stick again, and pushes the objective with great care towards the bars ; but even now she cannot avoid changes, although she always corrects herself immediately. Chica takes as little account as Sultan did at first of the length of her arm, and goes on pushing the objective towards the bars long after she could have reached it comfortably.

Next time, the solution is again worked out clearly from the very beginning. Chica does not once resort to the primitive direction, and indeed runs round to the bars before she has pushed the objective far enough towards the other side, thus showing, of course, that she has " calculated too favourably " her arm-length ; she returns once more to the gap, gives a few further pushes to the objective, and then completes the solution.

Two further repetitions on the following day result in clear procedures, with the exception of short tendencies to changes, which are corrected immediately.

i tried to test Rana also with such an arrangement, but sooⱶ had to give it up, as she seemed to consider it a point of honour not to deviate from 0° in any circumstances, and she could not be diverted by any sort of help, not even by repeated examples.

(Since the experiment just described has some similarity with that of the "roundabout-way board," it must be pointed out that it was undertaken *about a week later*. Sultan had already taken roundabout ways (on 18th and 19th March), at 180° ; Chica had failed when the board was in the normal position. The next board experiments took place a month and a half after the experiment just described.)

The experiments with the basket—mentioned in the Introduction—with rope, ring, branch stump, or nail, partly belong here, because in those, too, the animal had to find a solution at *one* place which would only work out as a solution at *another place*, after a detour. Further tests with that arrangement will be reported in a continuation of this book.

The board experiment and the tests described thereafter require, it is true, some adaptation of the movement to the given forms, but neither the things with which the proper roundabout ways are to be made, nor the structure of the field need be conceived with great precision as far as form is concerned, to achieve a solution. In fact, the solution was achieved on a very large free field.[1] If anyone wants to go on to higher demands, there is the exact fitting of one form which the animal is using, to another. Investigations along this line, which might be of the greatest importance for an understanding of the theoretical nature of intelligence, do not generally lead to very gratifying results with chimpanzees, and from our experiences up to date, failures and obscure behaviour are the only results we can expect in difficult cases of this sort.

(25.3.1914) Sultan tries to reach the objective placed behind the bars, with a stick, one end of which is bent into a crook. He seizes his tool by this crook, to poke it between the bars, and gets stuck with the crook caught behind one of the bars. This mishap leads to a hasty ramming against the obstruction, the factor of shape not being taken into account, and, when the stick is eventually freed, one has the impression that it was accidental. Some repetitions proceed in a similar way.

Two years later (in May 1916) the experiment is made with the same stick, in order to find out whether the animal is now capable of any greater clearness. As a matter of fact, Sultan, quite unmistakably, manages to keep the crook perpendicular, to fit the structure of the bars. He does this

---

[1] Only the experiment with the ring and the nail approximate to the following tasks in this respect.

while the crook is still far away from the bars, and so succeeds without difficulty ; in a few cases where he acts with less caution, and gets caught behind a bar, he quickly glances at the place of obstruction, and every time pulls the stick back and turns it, so that it can be got through without any further difficulty.   During this experiment the animal behaves more calmly than in the earlier one.

[Sultan seems not to notice or to realize the advantage which the crook offers, for instance, to pull a banana to him ; according to how he has picked up the stick, he puts the crook behind the objective, or else uses the point, as with every other stick.   Nueva, who from the very beginning, got the hook through the bars without trouble, may perhaps have recognized its advantage.]

Across one end of a stick about eighty centimetres long, a second stick of thirty centimetres is nailed, so as to form a " T."   Otherwise the task is the same as in the previous experiments.

(2. and 3.4.1914) Sultan endeavours to break off the cross-bar ; when this fails, he pushes the long part through the bars ; the cross-piece catches, and the animal rams it violently, blindly, and continually against the bars, until finally, obviously quite by chance, the cross-piece happens to be turned, so that it no longer sticks.   After about twenty repetitions, no observable improvement occurs ; it is evident that no attention is paid to shape.

Chica proceeds somewhat more quietly, but otherwise no better ; after a series of observations, I was forced to note that she did not even *attempt* to arrive at clearness.

Sultan is tested again with the same " T "-shape in 1916. As in the crooked-stick experiment, an essential improvement has occurred, inasmuch as the cross-piece, from the very beginning, while still at some distance from the bars, is held vertically, so that it can be got through the bars. *One has*

*the impression that Sultan is being taught by the forms which
he sees, what to do when stick and bars are opposite each other,
but not yet in optical contact.* As soon as, through carelessness
or haste, close optical contact of stick and bars has come about
(without a vertical turn in advance), Sultan's further procedure
depends on the special configuration in each case ; if the long
piece is in a position perpendicular to the surface of the bars,
while the cross-piece is stuck behind a bar, the latter is generally turned up with one sure movement, and thus put through ;
this is specially true in the cases where the twist required
forms a small angle (as was to be expected from previous
experiments). *When, on the other hand, the long piece itself
lies aslant, and the region round the junction of the two sticks
forms, with the bars, a relatively confused combination of lines,
then Sultan pulls and jerks blindly at his tool.* He is similarly
perplexed when he wants to get the whole thing in from
outside, and gets it muddled between the bars ; he then
simply tugs without paying any attention to shape. Not
every complex has the qualities of a good and precise
" *Gestalt* ", and even for the human being who looks on, the
cases that are not clear to Sultan constitute " less good
forms ", and therefore give no direct indication of the motions
required.

The forms which are to be handled in relation to other forms
are still more difficult to grasp visually. The ladder previously mentioned lies crosswise outside the bars and must be
brought in so as to reach an objective high up.

(12.5.1914) Grande and Chica seem to regard the task
as impossible ; in their discouragement they hardly touch
the ladder.

Sultan at first behaves in the same way. After some
time, however, he seizes the ladder, pulls one end across

between the bars, and tears it wildly towards the inside, although he cannot possibly achieve a solution thus. In his pullings and haulings the ladder finally goes somehow between the bars. In the course of a few repetitions the observer notices certain differences. Not every junction of ladder and bars is treated in an equally unintelligent way; on the other hand, some factors occur like those already mentioned in the previous experiment. Sultan does not know how to help himself out of that criss-cross mix-up which puzzled him also in the " T "-shape case; on the other hand, some of his twists are genuine, when the ladder is only a little out of the proper position. In general, then, the animal's movements are more or less clear according as the aspect of the lines of ladder and bars taken together is clear.

This experiment is repeated again later (in May 1916). The total impression of Sultan's behaviour is unfavourable as before; we cannot fail to recognize that intelligent behaviour alternates with quite absurd pulling and tearing, as the combination of forms of ladder and bars alternates from simple to complicated. But even for the human adult there are many cases and moments of optical " confusion ", although the observer, by a little effort, can always recover the required clearness.

The ladder experiment suggested to me that the test would be very much facilitated by introducing a solid form instead of the combination of lines used up to now (of ladder and " T "-shape). Therefore, the following situation is set up: the objective lies in a big box, and can only be reached from one opening, which is cut rectangularly in one of the sides (about ten by three centimetres); the objective, however, is so far away from this opening that a wooden board—the only stick in sight—must be used to get it. Its cross-section repeats on a small scale the rectangular opening, and, when

turned the right way, it can easily be put through the opening
into the box. (The animal can look inside the box through
other cracks, and the cross-section of the board is so much
smaller than the opening that this implement can easily be
managed.)

(6.4.1914) In this test both Sultan and Chica act without
much " order ". Both show that they are by no means
indifferent to the forms before them, for they soon turn the
board into the position *approximately* required before they get
to the opening. But, if there is the slightest check at one
corner, this failure does not have the effect of teaching them
to be more careful ; on the contrary, they push and jam the
board only the more wildly and blindly, until finally their
behaviour shows no regard whatever for the forms with which
they are dealing. (There are adult men who, in similar
situations—fighting with collar studs and so forth—behave
similarly ; the fault here lies more in the emotional field, in
character and " education ", than in the purely intellectual
field ; anyhow, from a practical standpoint the result is that
the processes proper to intelligent conduct no longer occur
in the degree otherwise possible, as soon as intense emotions
master the organism.)

I do not report further variations on the principle of these
experiments, as the results were always the same : with the
cleverest animals a clear adjustment to forms, as long as the
given structure remained clear and simple ; on the other hand,
even with the most gifted ones, a completely unintelligent
pulling and pushing as soon as forms were at all complicated.
After many experiences in this field it becomes more and more
certain that impatience and temper are not alone to be blamed
for this ; the same difference is also to be noted on the animals'
good days, and when they went quietly to work. The more
gifted chimpanzees show a certain improvement at the age
of about five to seven years, but the two older animals,

Tschego and Grande, are not ahead of Sultan (to say nothing of Nueva), in proportion to their age. As far as Nueva is concerned, we have to report certain data concerning her command of forms in another connexion.[1]

If the less gifted animals have not been mentioned very much in this section, it is to be attributed only to the fact that there is little to be said about this unintelligent treatment of forms, even those that are relatively simple ; the experiment with the detour-board may serve as characteristic of them all.

[1] See Appendix.

CONCLUSION

THE chimpanzees manifest intelligent behaviour of the general kind familiar in human beings. Not all their intelligent acts are externally similar to human acts, but under well-chosen experimental conditions, the type of intelligent conduct can always be traced. This applies, in spite of very important differences between one animal and another, even to the least gifted specimens of the species that have been observed here, and, therefore, must hold good for every member of the species, as long as it is not mentally deficient, in the pathological sense of the word. With this exception, which is presumably rare, the success of the intelligence tests in general will be more likely endangered by the person making the experiment than by the animal. One must learn and, if necessary, establish by preliminary observation, within what limits of difficulty and in what functions the chimpanzee *can possibly* show insight ; negative or confused results from complicated and accidentally-chosen test-material, have obviously no bearing upon the fundamental question, and, in general, the experimenter should recognize that every intelligence test is a test, not only of the creature examined, but also of the experimenter himself. I have said that to myself quite often, and yet I have remained uncertain whether the experiments I performed may be considered "satisfactory" in this respect ; without theoretical foundations, and in unknown territory, methodological mistakes may quite well have occurred ; anyone who continues this work will be able to prevent them more easily.

At any rate, this remains true : Chimpanzees not only stand out against the rest of the animal world by several

S

morphological and, in the narrower sense, physiological, characteristics, but they also show a type of behaviour which counts as specifically human. As yet we know little of their neighbours on the other side, but according to the little we do know, with the results of this report, it is not impossible that, in this region of experimental tasks, the anthropoid is nearer to man *in intelligence too*, than to many of the lower monkey-species[1]. So far, observations agree well with the theories of evolution ; in particular, the correlation between intelligence, and the development of the brain, is confirmed.

The positive result of the investigation needs a limiting determination. It is, indeed, confirmed by experiments of a somewhat different nature, which will be recounted later ; but a more complete picture will be formed when they are added, and, in so far, our judgment of the intelligence of apes is left some scope. Of much greater importance is the fact that the experiments in which we tested these animals brought them into situations in which all essential conditions were actually visible, and the solution could be achieved immediately. This method of experimentation is as well adapted to the chief problem of insight as are any which can bring about the decision " yes " or " no " ; in fact, it may be the very best method possible at present, as it yields very many, and very clear, results. But we must not forget that it is just in these experimental circumstances that certain factors hardly appear, or appear not at all, which are rightly considered to be of the greatest importance for *human* intelligence. We do not test at all, or rather only once in passing, how far the chimpanzee is influenced by factors not present, whether things " merely thought about " occupy him noticeably at all. And most closely connected with this, is the following problem.

[1] For reasons to be dealt with later, of course *not in applications* of intelligence. In this respect, no doubt on account of a general weakness in his whole organization, the chimpanzee is more nearly related to the lower monkeys than to man.

In the method adopted so far we have not been able to tell how far back and forward stretches the time " in which the chimpanzee lives " ; for we know that, though one can prove some effects of recognition and reproduction after considerable lapses of time—as is actually the case in anthropoids—this is not the same as " life for a longer space of time ".[1] A great many years spent with chimpanzees lead me to venture the opinion that, besides in the lack of speech, it is in the extremely narrow limits in *this* direction that the chief difference is to be found between anthropoids and even the most primitive human beings. The lack of an invaluable technical aid (speech) and a great limitation of those very important components of thought, so-called " images ", would thus constitute the causes that prevent the chimpanzee from attaining even the smallest beginnings of cultural development. With special reference to the second fact, the chimpanzee, who is easily puzzled by the simplest optical complications, will indeed fare badly in " image-life ", where even man has continually to be fighting against the running into one another, and melting together, of certain processes.

In the field of the experiments carried out here the insight of the chimpanzee shows itself to be principally determined by his optical apprehension of the situation ; at times he even starts solving problems from a too visual point of view, and in many cases in which the chimpanzee *stops* acting with insight, it may have been simply that the structure of the situation was too much for his visual grasp (relative " weakness of form perception "). It is therefore difficult to give a satisfactory explanation of all his performances, so long as no detailed theory of form (*Gestalt*) has been laid as a foundation. The need for such a theory will be felt the more, when one remembers that, in this field of intelligence, *solutions* showing insight necessarily are of the same nature as the structure of

[1] Cf. Appendix, p. 271, seqq.

the situations, in so far as they arise in dynamic processes *co-ordinated with* the situation.

One would like to have a standard for the achievements of intelligence described here by comparing with our experiments the performances of human beings (sick and well) and, above all, human children of different ages. As the results in this book have special reference to a particular method of testing and the special test-material of optically-given situations, the psychological facts established in human beings (especially children), under the same conditions, would have to be used. But such comparisons cannot be instituted, as, very much to the disadvantage of psychology, not even the most necessary facts of this sort have been ascertained. Preliminary experiments—some have been mentioned—have given me the impression that we are inclined to over-estimate the capabilities of children of all ages up to maturity, and even adults, who have had no special technical training in this type of performance. We are in a region of *terra incognita*. Educational psychology, engaged on the well-known quantitative tests for some time, has not yet been able to test how far normal, and how far mentally-deficient, children can go in certain situations. As experiments of this kind can be performed at the very tenderest age, and are certainly as scientifically valuable as the intelligence tests usually employed, it does not matter so much if they do not become immediately practicable for school and other uses. M. Wertheimer has been expressing this view for some years in his lectures ; in this place, where the lack of human standards makes itself so much felt, I should like to emphasize particularly the importance and—if the anthropoids do not deceive us—the fruitfulness of further work in this direction.

Postscript.—When I finished this book, I received from Mr. R. M. Yerkes (of Harvard University) his work entitled

The Mental Life of Monkeys and Apes: a Study in Ideational Behaviour (*Behaviour Monographs*, III, i, 1916). In this book some experiments of the type I have described are recorded. The anthropoid tested is an orang-utan, not a chimpanzee, but, as far as one can judge from the material given, the results agree with mine. Mr. Yerkes himself also thinks that insight must be attributed to the animal he tested.

APPENDIX

SOME CONTRIBUTIONS TO
THE PSYCHOLOGY OF CHIMPANZEES[1]

THE following report consists for the most part of " qualitative " observations on chimpanzees. The behaviour of
the anthropoid ape is in many essentials so important and so
easily comprehensible to man, that any minute experiments
in this direction are unnecessary for the present ; at the same
time the results of special experimental investigation only
take on the real colour of life when the habits and the character
of the creatures under observation have become adequately
known in their natural expression.

As a reaction against the assertions of marvels made by
inspired *dilettanti*, there has arisen among animal psychologists
a distinct negativist tendency, according to which it is considered particularly exact to establish *non*-performance, *non*-
human behaviour, mechanically-limited actions, and stupidity
to animals. Is not too much honour paid to the errors we are
combating by this negative attitude ?   Let us not, in avoiding
one error, be led to the opposite extreme. Unfortunately,
there are still those who, for various emotional reasons, wish
to find these or those qualities in the higher animals. For my
part, I have tried to be impartial, and I believe that my
description is not influenced by any emotional factor,

[1] Originally published in *Psychologische Forschung*, I, 1921.

beyond a deep interest in these remarkable products of nature.

I

I have already expressed the opinion that " the time in which the chimpanzee lives " is limited in past and future. First of all, the number of observations is small in which any reckoning upon a future contingency is recognizable, and it seems to me of theoretical importance that the clearest consideration of a future event occurs when the anticipated event is a planned act *of the animal itself*. In such a case it may really happen that an animal will spend considerable time in preparatory work quite unmistakably, as when Sultan labours long to sharpen one end of a wooden board, so that it will afterwards fit into a tube, and he can carry out his scheme with the double-stick.[1] Where such preliminary work, obviously undertaken with a view to the final goal, lasts a long time, but in itself affords no visible approach to that end, there we have the signs of at least some sense of future. To be sure, there is, in the example given,[2] the incentive of the visible reward, and, all through his labour, he could glance from time to time at the fruit. Anyone seeing an ape making preparations for an anticipated future experiment, the conditions for which are not at the time in sight, would be witness of a still higher achievement in the direction under discussion. Then the considering of certain *external* circumstances in the near or distant future, not only of self-planned actions, would operate as a condition of the actual behaviour. I have not yet made any clear observations of

---

[1] Cf. above, p. 125, seqq.
[2] Several others were mentioned.

this sort, nor, indeed, have I purposely arranged any situations suitable for them.[1]

Twice in the space of a few days experiments were carried out, in which the animals had to swing themselves, by the gymnastic rope, up to an objective hung high up on the wall, and when I, soon after, came upon the scene with my pockets bulging, one of the animals went through all the preliminary motions for swinging with the rope towards the goal, before I had taken any objective out of my pocket or begun my preparations for hanging it up. When frequent experiments were being made with boxes (as footstools) the chimpanzees would drag their implements eagerly to the accustomed test-place, if I merely showed myself at the door at the usual time. These cases are not quite clear : the appearance of the experimenter, that has usually been followed by certain occurrences, might be according to current opinion an immediately and directly reproductive stimulus for certain behaviour —but on the other hand the apes in no wise repeated old movements mechanically : their bearing displayed eager anticipation, and, finally, impatience, directed in short, and beyond any doubt, upon what was to come next.

Another experiment would make matters clearer : an ape that has often used boxes to reach an objective, is kept in a room where there are boxes at his disposal, but no objective for which to use them. His ration is cut short, but after a while he is taken into another room where there is plenty of food—

The determination of the question whether the chimpanzee can accomplish such a feat is important, I think, for the following reason : A number of the most various observations on anthropoid apes has shown us phenomena which usually appear only with beings possessing some culture, however piimitive. Since the chimpanzees have no culture worthy of the name, the question arises, which aie the limitations upon their capacities that are at fault ? Even the most primitive man makes ready his digging-stick, though he is not going to dig right away, and when the objective conditions for the use of the tool are not at hand. And the fact that he prepares thus in advance, is associated with the rise of culture. Hence I emphasize this question.

if only there were boxes with which to reach it. The way back into the first room is barred (else the case would be the same as that quoted on p. 52, seqq. above). After a while the hungry animal is let back into the first room, then again into the second, and so on, until a box in the first room might be seen as a tool for the situation in the second, and taken along for that purpose. (In observation it is a matter not only of the box being taken along, but of the whole behaviour of the animal towards the box in room 1, and especially of how the taking along *starts*.) This is the ground-plan of the scheme, the execution and the variation of which would develop automatically in practice.

To be sure, the chimpanzees had had, to start with, ample opportunity to adjust their actions to a future situation ; for example, when they faced the problem of learning to pick one among several, in one respect easily-distinguishable, objectives ("choice-boxes "). The chief difficulty for them was that at first they did not discover or see as essential the distinguishing properties or differences between these boxes[1]. Thus there were, at first, many mistaken choices, with the consequent disappointments ; therefore, when they lighted on a correct choice, the animals would have done well to look carefully, for future trials, at the right objective, and what distinguished it. As a matter of fact, one never saw them deliberately concentrate on the successful choice with an eye to the future ; on the contrary, the animals were carried away by their immediate and narrow interest in the goal before them (food), and if now and then a glance settled for a moment on the objects they had chosen from, this seemed to occur only because something chanced to strike them ; not intentionally in order to turn the lucky choice experience to future use.

Undoubtedly a ("voluntary") shifting of the attention

[1] *Nachweis einfacher Strukturfunktionem usw. Abhandl. d. preuss. Akad. d. Wissensch.*, 1918. Phys.-Math. Section, No. 2 p. 50 seqq.

from so strong a momentary interest, merely on account of the expectation of a greater general advantage later, would be a very notable achievement. If so much is not demanded, and if a somewhat more comprehensive behaviour brings at once a visible advantage, then inhibition of the most direct and primitive " urges " does take place. Most animal species cannot withstand the temptation of eating, even when it would be far more advisable to make sure first of the greatest possible supply.[1] It is not so with the chimpanzee. When I first began to feed all the apes together in one room out of a vessel, several of the animals began, without any outside interference, to postpone feeding altogether, or only hastily put a bite in their mouths in between whiles, as long as any was left in the vessel, or until they had a satisfactory amount stowed away safely in their hands, feet, and bulging cheeks. Meanwhile, they urgently begged or demanded an increase of their supply, and for the actual eating waited till they were in some quiet corner. Here it is most probably fear of competition from others which makes food as a " large quantity " for the time being more valuable than the satisfaction of their immediate appetites ; and when an animal locked up by itself behaves in the same way at feeding, it is probably the fear that the feeder might go away with the food that makes him do so.

I do not think that this conduct is to be traced to any consideration of the future properly speaking, as if the ape had said to himself : " If I don't grab as much as I can now, instead of eating it straight off, I shall not have enough later and shall have to go hungry." Even in the face of the following remarkable example, the assumption of so much looking into the future is altogether too intellectualistic. In the

---

[1] We are, of course, not concerned with those " collectors " in the animal kingdom, who from peculiar instincts store up nourishment as eagerly as they eat it.

evenings, the animals, in going from their stockade to their sleeping-dens, have to cross a space to which they have no access in the daytime, and which, therefore, in the rainy season usually becomes covered with weeds. They all storm upon this green food, which they are very fond of, and if they are left at it any length of time, it is very difficult to make them stop and go to their dens. We observed, over and over again, in two animals especially, that they took no notice of a first summons, at a second merely turned round carelessly, and at a third call, hurried on with their eating as hard as they could go. But when the warnings became more urgent, and we even advanced threateningly upon the miscreant, he would suddenly stop eating, tear up weeds with all speed, stand up and collect more of the choicest bits, go a long way round collecting all he could hold, and at last slip into his den with a tremendous bundle of the stuff. It is not necessary for an explanation at all—indeed, it is a very wrong hypothesis—to imagine the ape spurred on in all this by an " idea " of what it will be like in his den afterwards. As a matter of fact, we have here a case of roundabout behaviour, called forth by the circumstances, very similar to numerous ones related before. The goal to be attained, namely : " as much as possible of this lovely food ", is suddenly arrived at by an indirect, instead of by the biologically direct but obstructed path, and in those other tests also the animals will seldom have worked with an image of a later condition of affairs. The behaviour which solves the problem arises much more directly from a consideration of the present only.

[The word " future " is here used in its usual sense, so that " to make a picture of the future " refers to situations lying outside the actually occurring unit or whole of behaviour. It is only after one has conceived the present as a static " point "—against the nature of all psychological and physiological phenomena—that one can say that striving for a visible goal and fleeing from a danger actually threatening, imply a going beyond " the present " and a reaching into " the future." But this incorrect method of expression ought then to be applied to *every* case of emotional striving towards or away from things simply because in the unreal *present point* the dynamic essence of " drive behaviour " finds no place.]

The time in which a chimpanzee lives would reach very far back towards the *past*, if one were to apply as a sufficient criterion simply the plain after-effects of past experience in present actions. Perhaps no one will be surprised to hear that the animals knew me again at once, after a separation of six months (just as Sultan, after being away from the other animals and not having seen them for four months, was immediately upon his return hailed as comrade). For there are dogs who, after being separated from their masters for even longer periods, manifest the wildest joy on seeing them again. The fact that even after long lapses of time one finds the effect of an earlier learnt discrimination test as strong as though only a few days had elapsed, is perhaps more remarkable ; for the objects of which we are here speaking have nothing like the direct affect-value of familiar persons (or comrades).

Thirteen months after their last experiments in perception of size, Grande, in a first series of ten tests made *one* error (in the seventh test), while Chica chose all correctly.

Likewise thirteen months after her earlier period of learning, Tercera chose between two different reddish-blue colours all but one out of ten, exactly as before.

About eighteen months after the experiments on *Farbenkonstanz*, Sultan made *one* wrong choice out of the first ten tests, Grande no mistakes at all.

Quantitatively, therefore, the animals lost nothing ; their choices were merely made at first a little slowly and hesitatingly.

Without doubt, the interval could be very considerably increased, and there would still be a strong after-effect.[1]

For these achievements, there was certainly no necessity whatsoever for any image of the past. The familiar situation appears again immediately in the form of apprehension acquired in the learning period, and so brings about the same direction of choice behaviour. Almost the same thing will happen if, in the repetition of an intelligence test, the animals achieve the solution much more quickly than the first time, and if they show the same ability in the same situation even years later.

As long as memory works only in this way, it *may be* an advantageous gift ; but also, it may be a real hindrance to the appearing of valuable new behaviour, as I have before pointed out in connexion with crude examples furnished by the chimpanzees.[2] On the other hand, where real remembering broadens the scope of those conditions, which influence the life of the animal, this real extension of the life surveyed impresses the observer considerably : and " free ", " enlightened " does the ape then look, compared with the " narrowness of time " in which lower animals live.

Whether animals " have *images* " (in distinction to all *perceptions* altered or not altered by experience of their earlier life), is a question with which American animal psychologists since Thorndike have occupied themselves considerably. Hunter first made experiments which involve something like real " remembering," rather than the reproduction of former behaviour in a like situation. The animals he experimented on first of all had to learn to choose the one out of three open entrances, which was, each respective time, illuminated. When they satisfied this condition, they were allowed to approach the entrances only after the light had vanished ;

---

[1] Cf. *Nachweis einfacher Strukturfunktionem, etc.*, p. 78.
[2] Cf. above, p. 195.

i.e., after seeing this signal, they had to wait a certain time, opposite the three doors now looking alike (delayed reaction). The longest interval after which decisions were still correct was, in rats, ten seconds, in dogs, five minutes, and then they only chose correctly because they turned towards the light while it was still there, stood fast, and afterwards ran straight ahead, so that if they *turned* at all in the interval, they made the mistake corresponding to that turn. A racoon selected the correct entrance even when he had turned his body as much as he liked in the interval, but he did not manage it when the interval was longer than half a minute.[1]

One sees at once that we have here a great difference between animals and man ; in fact, children between the ages of six and eight, in a similar case, did not reach their limit even in half an hour, and only a little mite of two and a half could not choose after one minute. Buytendyk rightly calls attention to the fact that, in Hunter's method, the preliminary training to understand the " light business " causes unnecessary and unnatural complications ; in quite a simple qualitative test he shows that a monkey (Cercopithecus) could carry out the delayed reaction with certainty even after seven minutes.[2]

I made the following experiments on chimpanzees :

1. Sultan is sitting alone in a barred space which contains no sticks. Outside, in a homogeneous dry sandy place before his eyes, 1.40 metres away from the bars, I bury a pear some centimetres deep, and I wipe out every trace of the hole by smoothing evenly over the spot and all round it, so that I could not recognize the place myself. Sultan, who first looks very disappointed, soon begins to play, apparently showing no further interest in what he has seen. When I approach

---

[1] *The Delayed Reaction in Animals and Children. Behaviour Monographs*, 2.1.1913.
[2] *Arch. neerland. de physiol. de l'homme et des animaux*, 5, 1920.

him after about six minutes, he quickly seizes my hand, as if
to lead or pull me, but he is repulsed.[1] Nine minutes later,
when I again approach, he repeats this at once, and it is now
clear that the animal is trying to drag me to a stick, which is
lying outside, some distance from the burying-place, and
which cannot be reached from the bars. As I will not give
in, Sultan stops and keeps quiet until (seventeen minutes
later) I come near to him again and he can repeat his attempt.
Just half an hour after the beginning of the experiment while
he is busy with something else and is not looking out, the
stick is put close enough to the bars for him to be just able
to reach it with some effort. At first Sultan does not notice
this change ; but when, however, his glance again happens to
fall on the stick, he springs up, pulls it in, runs quickly with
it to the bars opposite the burying-place, and scrapes the
sand away at the exact spot, until the pear appears. The
railings through which he is working are five metres long,
and he can reach out with his stick about two metres. From
this test we deduce the *keenness* of his memory (considering
the absolutely uniform surface) ; its liveliness is expressed
in the promptness of his action, which he carried out as soon
as he could. Of course, it goes without saying that the
experimenter did not pay any attention to the burying-place
in the meantime.

2. On the following day I tested Sultan again. I buried
a pear 1.30 metres from the bars, but two metres to the side
of the old burying-place, and spread a uniform layer of sand
over it and the surrounding area. There was no stick within
reach, and none to be seen. Sultan this time remained
quietly playing in his den. Exactly an hour later, during
which nothing had happened in connexion with the buried
pear, the keeper threw a stick through the window at the
back of the cage. Sultan picked it up at once, ran to the

[1] For similar experiences, of. above, p. 142 seqq.

bars, just opposite the right place, and scraped the sand away about thirty centimetres too far from the bars ; but after a moment, he put the stick nearer, this time exactly on the spot, and quickly brought out the treasure. This time he beat me, for after his shift I thought he had missed the spot, and it was not until the fruit appeared that I was undeceived. I wish to stress the fact, that the animal had paid no attention whatever to the burying-place in the interval, but had climbed about unconcernedly in his cage, and, indeed, had more often turned his back than his face to the bars.

3. Three days later, towards evening, a heap of fruit was buried in quite another part of the animal's stockade, but in the presence of the apes, who were attentively looking on. Immediately afterwards, the animals went to rest in their dens, and when all was quiet, we carefully smoothed over all sign of the burying-place, and then dug some holes a few metres away, which we left empty, but covered over, like the first. The next morning (sixteen and a half hours later) the apes were let into the same place again. When I opened the door for Sultan, he tossed me a comparatively short " good morning ", and hurried past me in a straight line to a place about sixty centimetres from the right one ; he began to scrape away the sand, but soon stopped, when he came to hard ground. The others tried the same thing on the same spot, but they also stopped, and only after a pause was the right place suddenly discovered. In this test, not one of the controls was touched. The field of possible error was at least four hundred square metres in extent, but was more marked by some posts, and by vegetation here and there, than the sandy surface of the first experiment.

4. The same preparations were again repeated in another place, four days later, in the presence of the animals. And after a night (sixteen and a half hours in all) we conducted the test again, Sultan, however, taking no part. The oldest

T

of the animals left his cage and squatted somewhere near, without any sign of remembering. But Grande, on the contrary, the second to come out of her sleeping-den, ran confidently and unhesitatingly straight to the right spot, and dug out the fruit.[1]

An extension of this test to longer periods, and, above all a closer analysis of the proceedings, are both very desirable. When experiments are made on testing-fields distinguished by marks of varying distinctness and on various principles, we may look with assurance for light upon the nature of the animals' perception, and memory. One can see at once how, through influencing the apes in the interval (diverting and misleading tests), and also, above all, through secret alterations in the visible structure of the space (before testing) we should learn more.

II

It is hardly an exaggeration to say that a chimpanzee kept in solitude is not a real chimpanzee at all. That certain special characteristic qualities of this species of animal only appear when they are in a group, is simply because the behaviour of his comrades constitutes for each individual the only adequate incentive for bringing about a great variety of essential forms of behaviour. Furthermore, the observation of many peculiarities of the chimpanzee will only be clearly *intelligible* when the behaviour and counter-behaviour of the individuals in the group are considered as a whole. In

---

[1] That the outcome of these tests with chimpanzees cannot be attributed to the sense of smell needs no further elaboration. Anthropoids have this sense to a very limited degree ; a hidden pear, tomato, or the like (the kind of fruit we used), is not smelt a couple of decimetres away. And yet in the last test the animals ran immediately and straight to practically the correct spot, over a distance of more than 10 metres.

this, the part played by one animal may have definite significance for the observer, which it would not have if a human being, for instance, were the (necessarily bad) partner. But, apart from this, the group connexion of chimpanzees is a very real force, of sometimes astonishing degree. This can be clearly seen in any attempt to take one animal out of a group which is well established as a group. When such a thing has never happened before, or not for a long time, the first and only desire of the separated creature is to get back to his group. Very small animals are naturally extremely frightened, and show their fear to such a degree, that one simply has not the heart to keep them apart any longer. Bigger animals, who do not show signs of actual fear, cry and scream and rage against the walls of their room, and, if they see anything like a way back, they will risk their very lives to get back to the group. Even after they are quite exhausted from these outbursts δf despair, they will crouch, whimpering, in a corner, until they have recovered sufficient strength to renew their raging[1]. As in tests with chimpanzees, appetite is a necessary condition, for the first few days after the isolation of the animal it is impossible to conduct any experiments, because they refuse food altogether at first ; and for a long while, even if hunger drives them to take a bite or two, they will immediately drop the food again apathetically.

Generally the remainder of the group, even when they hear his moanings from a distance, do not take the same strong interest in him, nor are as sad at the separation as he is. The others are still a " group ". One cannot say that they listen to his wailings without any sympathy. It often happens

[1] I have before described how the urge to get back to the others often leads to those extremely peculiar activities which, partly method of expression, and partly use of implements, show clearly that the creature isolated *must* do something in the direction of his feelings, even if it is of no practical help. Cf. above, p. 88, seqq.

that if it is possible for them to get near the prisoner's cage, one or other of the animals will rush to it and put his arms round him through the bars.   But he had to howl and cry for this affection to be shown him ; as soon as he is quiet, the rest of them do not worry ; they show no desire to get to him, and even his good friends soon stop embracing him, in order to return peacefully to the more important group.   It must not be imagined that the isolated ape is sad only because he is in a cage, and the others have more freedom.   For if one of them is outside, and the others in the cage, the one outside tries his utmost to get to the others in the cage.

[Exactly the same thing happens when an animal, who has been isolated from his group for some weeks for testing purposes, comes back again.   His joy reaches such a pitch that it may easily be accompanied by faint glottal cramp.   The others do not get *quite* so excited ; but as his return represents an actually impressive fact, the whole group becomes very lively ; they put their arms round him, even beat him a little for pleasure, and often the whole bunch run along behind their returned companion, as is the habit of chimpanzees, in order to examine minutely his rump and sexual parts.   At the same time, it matters a great deal who the returned ape is.   The oldest animal who occupied a special position in the life of the community, was, on any such occasion, greeted by a universal welcome, such as was not accorded to the others.]

More than once I established that the temporary (or permanent) disappearance of a sick (or dying) animal has little effect on the rest, so long as he is taken out of sight and does not show his distress in loud groans of pain, as, indeed, chimpanzees so rarely do.[1]   This corresponds to the lack of concern

[1] Sick chimpanzees soon grow quite apathetic and then complain as little about their isolation as, curiously enough, about the great pain which one can plainly see them enduring.   Is it not astounding that just the chimpanzee who, for the most trivial reason, becomes quite uncontrolled, giving vent to savage fury, hardly even moans softly when he is writhing on the floor with agony ?

of the group in the healthy ape that is segregated, as long as he does not whine too miserably ; and if a sick animal dies isolated in its own room, it is no use expecting any sign of sadness or of missing him, as there is no actually perceivable incitement to mourning or excitement, and every animal in the group at the moment feels the group around him. Unquestionably, their interest to-day in some fruit which they saw buried yesterday, is greater than that taken in one member of the group who was there yesterday and who to-day does not come out of his room any more.

But just as considerable—though transitory—interest is shown, when an isolated creature's wailings can be heard or seen, so also I noticed the strong effect on the others, when they once saw with their own eyes the signs of weakness and illness in one of the little chimpanzees. At the beginning of his fatal illness Konsul was once lying helpless on the floor, with his eyes closed. Rana, who happened to be passing by, asked him in the usual way to accompany her, as I have described it before.[1] As he hardly moved, and immediately sank back again, she grew attentive, first lifted his head, and then, putting her arms around the little fellow, carefully lifted his weak body, and seemed by her bearing and her look so deeply concerned, that there could be, at this moment, no doubt whatsoever as to the state of her feelings. When some days later, during which he had been isolated, she again saw the poor creature in a very wretched state, she seemed only to shy away. But again, one day when he seemed a little better, the little fellow was once more let out into the open, where the others were gaily eating green stuff. He dragged himself painfully to them, but after taking a few steps he suddenly fell to the ground with a piercing cry of fear. Tercera was sitting some way away, chewing. She sprang up, her hair standing on end all over her body with excitement. She reached him

[1] Cf. above, p. 49.

in a few strides, walking upright, her face filled with the utmost concern, her lips protruding with sorrow, and uttering sounds of distress ; she caught hold of him under the arms, and did her best to raise him. One could not imagine anything more maternal than this female chimpanzee's behaviour, and I give these words their literal meaning, as applied to conduct which was called forth at that moment by the immediate and strong impression made on her by the break-down she witnessed. The fact that Konsul, after being taken back to his room, never came out again, evoked as little sign of grief from Tercera as from the other members of the group. Therefore, when we compare the former conduct with the naturally ethical striving and behaviour of humans, we must not lose sight of the fact that, as an indispensable condition, the breaking down in weakness and helplessness has to be perceived as an impression phenomenon. The ape will feel no sympathy while merely imagining such a case, and naturally because he hardly ever has such images.

[Moreover, we should not deceive ourselves with too idealistic and unreal optimistic pictures about human beings in like circumstances. Here and there will be found a case of a sick person, not likely to get better, who is finally considered a burden even by his near relatives ; they almost get angry with him.]

If one chimpanzee is *attacked* before the eyes of the group, great excitement goes through the whole group. It will happen that, under the influence of the climate, one punishes a wrong-doer with a heavy blow. The moment one's hand falls on him, the whole group sets up a howl, as if with one voice. The excitement thus expressed has usually nothing of fear in it, and the group does not run away. On the contrary, though they are separated by the railings, they tend to approach the place of punishment. Even the lightest form of punishment, pulling the ear of the offender, or a playful pretence at punishment, often stirred single members of the group to much more decisive action. It was, in particular, little weak Konsul, who would run up excitedly, and, in the way little chimpanzees have of expressing their wishes, with a pleading countenance, stretch out his arm to the punisher, if the ape was still being punished, try to hold one's arm tight, and finally, with exasperated gestures, start hitting out at the big man ! Those who think it degrading to human beings to find anything human in animals, make the extraordinary assertion that animals never defend each other, and conduct, which may appear to be defence of another, is produced only by the fact that the foolish creature assumes itself attacked, and in this error goes to its *own* defence. But little Konsul would come running up of his own accord, although he had been quietly squatting a considerable distance away, in order to take part in such an occurrence ; and once even, when he was in another place, where he could not see what was happening, but only hear something of it, he hurried at once by a round-about way, and fell on my arm. Unquestionably the animal " was affected " by what was happening to his companion, but it was impossible for him to assume himself in danger.

[Is it still necessary to show that such things are of quite usual occurrence much lower down the scale in the animal kingdom? Every decent cock will come hurrying along when a hen of the farmyard has been caught and is heard cackling with fear ; he does not actually attack, but at the distance which the fear of man counsels him to keep, he gets so excited and furious, that his *inclination* to attack is quite obvious.]

When the apes have grown much older and their awe of us big humans has diminished, and especially after they have arrived at sex maturity, I find the drive of the group to repulse an assault on one of its members grown inordinately stronger. In the end, one has to give up punishing even bad offences, when the whole group is in the same room as the wrong-doer. At times the most insignificant episode between man and an ape, which arouses the ape to cry in anger against the enemy and spring against him, is sufficient for a wave of fury to go through the troop ; from all sides they hurry to a joint attack. In the sudden transfer of the cry of fury to all the animals, whereby they seem to incite one another to ever more violent raving, there is a demonic strength, coming, surely, from the very roots of the organism. It is strange how convincing, one might say full of moral indignation, this howling of the attacking sounds to the ears of man ; the only pity is that every little misunderstanding will call it forth as much as any real assault ; the whole group will get into a state of blind fury, even when the majority of its members have seen nothing of what caused the first cry, and have no notion of what it is all about. The only thing necessary to the uproar is that that scream shall be uttered in that characteristic manner that whips up all the others. It has happened to me that in such circumstances Rana, the good-tempered, has suddenly lost her head, and in a mad fury sprung at my neck, when the moment before she was playing happily with me.

It is part of the extraordinary variation in character of the chimpanzees that many of them will not intentionally incite to mass attack, while others, when they are in a bad mood, will readily do so, fly into a rage over a trifle, and behave viciously in order to incite the herd. This is unhappily the case with the gifted Sultan, whose disposition to drop into the rôle of the wronged and to be pitied, has been mentioned before.[1] Thus, in a fit of anger, with which the innocent observer has had nothing to do, he will attack him with fury. He hops, choking with his glottal cramps, and screaming, up to an older animal that has often helped him, whines, springs shrieking

---

[1] *Optische Untersuchungen usw. Abhand. d. Preuss. Akad. d. Wiss.,* 1915. *Phys.-Math. Section,* No. 3, p. 13.

back at the human, and so on, in a manner that is an expression of challenge, if behaviour ever expresses anything. It is commonly said that all emotional behaviour is " directed towards an object," meaning that it is that object that calls forth the feeling. There occur, however—among human beings too —emotional states which have at first no such object (the sullen moods of neurotics in the morning, for example), but which look eagerly round for and usually find a convenient victim. In the tropics the most extraordinary phenomena of this sort can be observed and indeed experienced in oneself (tropical choler). On the other hand, a very strong feeling, such as anger, tends, when checked from expending itself on the object that aroused it, to turn and expend itself upon an entirely different object. When Sultan was quite young, and I punished him, he, not daring to avenge himself upon me, would run in a fury at Chica, whom he could not abide anyhow, and persecute her, although she had absolutely nothing to do with the cause of his rage.

The limit of the " outside," against which the group as a whole reacts so strongly on specially impressive, affective occasions, is by no means determined zoologically ; the group is a vaguely-organized community of chimpanzees *used to each other*. One day a newly-bought chimpanzee arrived, and at first was put for purposes of sanitary control in a special cage a few metres away from the others[1]. She at once aroused the greatest interest on the part of the older animals, who

[1]The coming of this new member suggested a test. I did not know whether it was a male or female, and it happened that it arrived in a small box, through the window of which one could recognize nothing but its head. The whole character of its face immediately gave me the impression that it was a *female* ape. One after another I now allowed five people (of whom three were wholly or almost illiterate, but who knew the whole band of apes at the Station very well) to go to the box, and each one, independently of the others, by looking at its face only, to judge the sex of the creature. All the opinions were : " female ", and that was right. Even now, after many years' acquaintance with chimpanzees, I could not give a morphological, distinctive mark (in the sense of some *special* characteristic of head or face) by which to distinguish the two sexes in young chimpanzees.

tried their best with sticks and stalks put through the bars to indicate at least a not too friendly connexion with her ; once even a stone was thrown against the wire-netting at the new-comer, and any active proceedings taking place between us and the new arrival were accompanied by excited noises from the others. When the new-comer, after some weeks, was allowed into the large animals' ground in the presence of the older animals, they stood for a second in stony silence. But hardly had they followed her few uncertain steps with staring eyes, when Rana, a foolish but otherwise harmless animal, uttered the ape cry of indignant fury, which was at once taken up by all the others in frenzied excitement. The next moment the new-comer had disappeared under a raging crowd of assailants, who dug their teeth into her skin, and who were only kept off by our most determined interference while we remained. Even after several days the eldest and most dangerous of the creatures tried over and over again to steal up to the stranger while we were present, and ill-treated her cruelly when we did not notice in time. She was a poor, weak creature, who at no time showed the slightest wish for a fight, and there was really nothing to arouse their anger, except that she was a stranger[1]. On analogous cases in the poultry-yard, and, on the other hand, in the behaviour of many primitive groups of human beings, there is no need for me to expatiate further.

In the transition to gradual tolerance the group became a little less disorganized. Sultan, who had played less part in the above-mentioned assault, was the first to be left alone with the newly-arrived female. He at once began to busy himself with her in his most diligent manner, but she was really very shy after her bad treatment. However, he went on

---

[1] When little Koko was a new-comer he was received with such a tumult of screams of indignation through the bars, that I never dared to let him in with the others.

trying to make friends, with sparkling eyes and a most friendly manner, until at last she gave way to his invitations to play, to his embraces, and—rather shyly—to his childish sexual advances. When the others came near and he was any distance away, she called him anxiously to her ; and really he defended her most gallantly when any other member of the group advanced with inimical bearing. Whenever she was frightened, she and Sultan at once put their arms round each other. Two other female apes, however, likewise soon broke away from the muttering group, played with the new-comer and kept on putting their arms round her ; until at last only Chica and Grande, who up till now had shown no special friendship for each other, united by a mutual aversion, formed a conservative alliance and led their own life in distant spots of the stockade, away from the new-comer and the renegades. It will be readily understood that the new little ape, as long as she was frightened of the group, preferred us humans, who treated her well, to all the chimpanzees, even to those who tried to gain her friendship. When, after that first assault, the older animals whom we had chased away stood ready for a renewed attack, the wounded ape collected herself, tumbled forward with all haste towards the nearest of us humans, and clambered up him. At first it was quite impossible to keep her on the ground ; forced down, she climbed upon someone else, put her arm round his neck, wailing, while she excitedly stroked his back with one hand. And later on, it was difficult to leave her alone in the stockade, for, as soon as we walked towards the exit, she would come running after us, wailing and lamenting, and clamber up agitatedly, stroking with one hand and holding fast with the other three extremities.

As the limits of the group are not determined only zoo-logically, the behaviour of the animals belonging to the group, towards friendly human beings, may be very similar to their conduct towards a member of the group. The reader may have seen that from some previous remarks.   In this respect, what struck me most was that the animals were occasionally on my side.   Once when people came inquisitively round the station enclosure, to whom the animals had paid no attention before, I told them firmly that they must go away, but in vain.   The animals began to pay attention ; I called to the people sharply, and the whole group of apes shrieked with indignation.   I shouted at the hesitating intruders, and all the chimpanzees jumped howling against the bars at them. On very rare occasions, chiefly in the animals' younger days, it happened (in contrast to the behaviour above described) that one of them sided with me against his companions. Sultan who, on account of his better understanding and his gifts was the first to come closer to us, several times when I scolded an animal would leap angrily at him ; if he could not reach him while we were having our " tiff ", he sometimes attacked him afterwards.   To be sure, it must remain un- decided whether preference for me was the reason of this behaviour or that ugly and very common delight of people who, fortunately innocent in the actual case, enjoy their moral indignation against a wrong-doer who has been found out.   For in humans, too, such phenomena should not be considered as the outcome of very complex and intellectual processes ; they seem to be direct expressions of very simple dynamics of emotional behaviour, and so one may acknowledge these phenomena more easily than many others in the higher animals, especially since, in this case, Sultan is rather " that sort of character ".

[The following incident requires a different explanation.   I follow an animal which has misbehaved, and it flies screaming in among the others, and several times past the older ape, Tschego, who is staring straight in front of her, peevishly. As this excited performance is somewhat drawn out, Tschego, just at the moment when the hunted creature is rushing past, jumps up crossly, seizes one of his hands, and gives it a good bite. All her life, *peace* was Tschego's essential need. When a noisy quarrel broke out among the other animals and came near her, she always grew angry, sprang up, stamped her foot, and struck out with her arms at the disturbers of her peace ; if one of them came too near her, he was treated as in the case I have just described.[1]]

Even very astonishing conduct towards man can yet be quite unequivocal. When I had been in Tenerife a few weeks only, I noticed, whilst feeding the squatting animals, pressed up close to me, that a little female, at other times quite well-behaved, was snatching the food out of the hand of a weaker animal, and as she persisted in this, I gave her a little rap. The little creature, which I had punished for the first time, shrank back, uttered one or two heart-broken wails, as she stared at me horror-struck, while her lips were pouted more than ever. The next moment she had flung her arms round my neck, quite beside herself, and was only comforted by degrees, when I stroked her. This need, here expressed, for forgiveness, is a phenomenon frequently to be observed in the emotional life of young chimpanzees. Even animals, who when they have been punished, at first boil with rage, throw one glances full of hate, and will not take a mouthful of food from a human being : when one comes again after a time will press up close, with eager bearing, to which a quick rhythmic breathing and pulling open of the eyes belong, or also with a sob of relief, pressing one's fingers affectionately between their lips and making all other protests of friendship.

[1] The naïve chum-like manner of the animals towards us sometimes expressed itself surprisingly in Tschego. She was too dangerous to be punished otherwise than by a stone cautiously thrown from afar. After this had happened several times, Tschego, as soon as one of us was scolding her and at the same time picking up a stone to throw, would run to that person, take hold of his hand and hold it tight, without showing any special excitement. Then she would coolly take the stone out of his hand, throw it on the ground, and walk quietly away.

In such scenes it has been noticed time and again, that highly impulsive behaviour is not regularly directed towards concrete material *advantage*. Why is one *greeted* by the animals at all, and often more impetuously than one would wish, when one sees them for the first time in the morning? Is it only because their breakfast follows close on one's arrival? If this were so, then the embracing, etc., would mean *joy* at what was expected and the need of *imparting* it, and should not be interpreted universally as an attempt to get their food more quickly. It is often, indeed, just when the chimpanzees have been most impatiently awaiting their meal that they will at great length hail the person who is bringing their food with loud cries of joy, put their arms round him, slap him and each other with pleasure, pull him and his food-vessel hither and thither, until at last one after the other stop their excited cries of joy, and, after the self-appointed delay, quickly snatch a good bit. If the animals were merely in a hurry for their food, and this were at that moment their main object, they would act quite differently.[1] Therefore the hearty greeting, especially on seeing one again in the mornings, is by no means to be understood merely as rejoicing for the coming food either. The animals are pleased simply at seeing again the human friend, just as they are pleased at seeing each other again ; it is the same as with dogs. It is true that one becomes a " friend " if one feeds the animals regularly ; but in Tenerife the person who finally took over all the feeding almost without exception, was never treated to such hearty and extended greetings as the person who managed them better in play and in " business ".

I will give another striking example, to show how the momentary practical advantage can quite recede into the background, when some emotional state has first to be expressed. One night, when it was raining unusually heavily

[1] Cf. the description above, p. 143.

and continuously, I heard two animals, who were kept alone in a special place, complaining bitterly. Rushing out, I found that the keeper had left them out in the open, because he had broken the key to the door of that place. I tried to force open the lock. It gave way, and I stood aside to let the two chimpanzees run in as quickly as possible, into their warm, dry sleeping-den. But, although the cold water was streaming down their shivering bodies on all sides, and although they had just shown the greatest misery and impatience, and I myself was standing in the middle of the pouring torrent, before slipping into their den they turned to me and put their arms round me, one round my body, the other round my knees, in a frenzy of joy. And it was not until they were satisfied in this way, that they threw themselves into the warm straw of their sleeping apartment.

The incidents just described are characteristic of the behaviour of chimpanzees towards familiar *adult* human beings. With little children and infants there is no need for these apes to know them at all, in order to be taken with them. When a young creature like this was brought near the railings of the animals' play-ground, one or other of the animals regularly came up and looked at the apparition for a long time with interest, a good-natured, satisfied expression on its face, tried to look under the clothes encasing the infant and nodded from time to time pleasantly in the direction of the child in front of him. This behaviour was most noticeable in the eldest of the females. As she had only come to the station when fully-grown, it may be that she had already had dealings with baby chimpanzees. But much younger animals behaved in a similar way, and I found it very clearly marked in a female orang, long before sexual maturity.

[When an animal is ill, in the apathy and lack of energy which even minor ailments bring with them, there soon arises an unusual docility and need of support from the human who is tending him. Hardly is the whole tone of the body normal again and strength and independence returned, than this closer understanding diminishes.

I will deal briefly with a few more points regarding their characteristic attitude toward humans. If one is specially friendly to, and plays more, with any one of the animals, the others not seldom become jealous. For instance, when Tercera saw anything like this, she would begin to walk about restlessly, looking at me, and after piteous and reproachful sounds, she would approach me and nudge me again and again, so as to turn my attention away from the other animal to her, or else, pouting all the time, would try to push the other animal away and take its place. I must say that sometimes Tercera's behaviour came pretty close to that of a coquette.—When one tries to make an ape do something which he does not feel like doing, the effect of this pressure, as a rule, is merely that the greatest opposition is offered against doing what is wanted, just as, while it is quite possible to pull a chimpanzee's arm in play, the moment the pulling in any way appears as a restriction of his liberty, the muscles are at once tightened in the opposite direction. None of the animals were as obstinate as Sultan, and his behaviour, when I wanted to force him to perform a test, was like that of a wayward child. One day, when he made his choices between two objects too lazily, and I applied pressure, I found it impossible to make him do anything any more, not even to take in his hand the stick with which he was to choose. The other animals were fed, but not Sultan, and still he would not touch the stick, although he would have been able to get food with it immediately by choosing the right object. I put the others to bed and still Sultan remained stiff-necked. Then from a hiding-place I noticed the following : when evening came and it grew colder and less comfortable, he at last took the stick, scraped around the ground in his room with it, but exactly in the opposite direction from where the test-objects were. After a time he pushed it through the bars, and scratched about sideways in the sand, as though he were playing. Then he dropped the stick again, but after a few moments picked it up, and so it went on, until at last he made the simple choice and the right one, which I had demanded of him. When now I brought him into his sleeping-den, there was a frenzied scene of reconciliation, such as I have before described.—The same animal behaves quite extraordinarily, when one tries to teach him something in which he is not interested. Once, he was required, in the evening after all the animals had been fed, to collect the fruit skins which were lying about, and put them in a basket. He quickly grasped what was required of him, and did it— but only for two days. On the third day he had to be told every moment to go on ; on the fourth, he had to be ordered from one banana skin to the next, and on the fifth and following days his limbs had to be moved for every movement, seizing, picking up, walking, holding the skins over the basket, letting them drop, and so on, because they stopped dead at whatever place they had come to, or to which they had been led. The animal behaved like a run-down clock, or like certain types of mentally-deficient persons, in whom similar things occur. It was impossible ever to restore the matter-of-fact ease with which the task had at first been accomplished.

Never could I achieve educational effects lasting beyond the time of my actual presence by frequently repeating pro-hibitions and by punishment. If the chimpanzees have just been forcibly prevented from some activity which they like, but which has been forbidden them, and if one then hides oneself in order to see what will happen, it is very amusing to observe how they first look carefully all around in every suspicious place, and then, seeing no actual danger, gradually work up nearer to the place of the forbidden act, in order, soon, to begin most zealously to sin again, as though there were no such thing as a human being and no possibility of future reckoning. However, it is not only the actual punishment which afterwards frightens them. Their habit of eating their excrements was often, and finally very sharply, punished, but all to no purpose. Often, however, in going to their stockade, I would miss an animal, and after a search would find it in some box or other, or crouching on the floor behind some vegetables, its whole face smeared over with the traces of its odious meal. My approach sufficed to make it feel fear for what it had just done. At times the animals are sufficiently naïve to give themselves away by their restlessness, when you yourself come to them quite unsuspecting. Thus Chica once began to hop agitatedly from one foot to the other, when I happened to come up unexpectedly; I had noticed nothing unusual. As I got nearer, her agitation increased, and all at once she let a mess of rubbish fall out of her mouth. More striking was it when one day Chica received me with the same disquieted hopping and would not stop, although I could not discover any guilt in her. Thus made attentive, I became aware that her friend Tercera was missing, or rather that a piece of her black fur kept disappearing behind a box each time I came round the other side of it. Nearer investigation showed clearly that this time *she* was the sinner. Since one animal will often plead for another who is going to be punished, Chica's behaviour on this occasion was after all quite comprehensible.]

The coherence of the group is by no means homogeneous as between all the members. In Tenerife any animal that distinguished itself in any way played a special social rôle for the rest. Tschego who, as the oldest and strongest member of the group, and commanding the most respect, was the one to whom the rest ran in time of danger, and whose support each party tried to win when there was a quarrel, easily carried the whole troop with her when she changed her occupation or place. But there was Rana also who, on account of her stupidity and her dependent, dull behaviour, was for the most part *de trop*, which state of things she did not improve by perpetually trying to approach the others ; on the contrary, she only made herself the butt of all kinds of practical jokes.[1] Secondly, there are in the relations of any two animals all grades of friendship and even qualitative colourings down to a small dislike, which apparently agrees well with the pervading social union of the whole group. Some of these special relations lasted all the time I made observations on the group, or as long as the animal concerned lived. Rana, rejected over and over again by the bigger animals, had taken possession of little Konsul, and never tired of him till his death. Tschego and Grande were all the time a little group in themselves within the larger group, and the friendship between Chica and Tercera lasted through all the changes of time, Tercera always remaining the strong, helpful, " giving " half. In the course of everyday life, these old predilections might almost escape notice ; but it only required fear or danger for them to be at once expressed, in seeing who embraced whom, and which two retired into a corner together. Also in the spontaneous distribution for sleeping these proved friendships were adhered to ; because the younger chimpanzees prefer to sleep in couples the whole night through, with their arms around each other.

In less important situations it is easy to overlook these

---

[1] The marked leadership of the little male Sultan, which Rothmann and Teuber (*Abhandl. d. Kgl. Preuss. Akad. d. Wissensch.*, 1915, *Phys.-Math. Section, No.* 2), thought they recognized very early, proved to be an artificial product. The animal was naturally favoured by his better understanding, and this made him stand out from the group. But I noticed that he bore his distinctions badly, and took steps accordingly, and soon there was no more to be seen of his leadership in the group. It was not until he was quite grown-up that Sultan really became master.

strongly cemented relationships, because they are often overlaid by weaker and constantly changing friendships. Rana had to give up her Konsul to all the other animals in turn, because each of them, at one time or another, took a special fancy to him. Big Tschego, at first, had quite a special penchant for the older little male, Sultan, who enviously tried to keep this preference on the part of the head of the group for himself, by attacking any other animal who dared to approach. But after his character had earned for him one or two hard reproofs from Tschego's hand, he quite lost his rôle of confidant later on, and it looked most comical to see him, with increased respect and slightly retiring, squatting near her, but quite unnoticed; how he would scratch his head with a more and more disquieted expression on his face, at the same time still trying to chase the others away from Tschego, until at last she herself got angry, and drove him away.

The particular degree of friendship existing at any time has a special significance, when there is a question whether a chimpanzee, who is begging for food from another, will get it or not. According to current opinion such a thing never occurs, on the contrary, one animal definitely envies another his food. And, as a matter of fact, one will wait in vain for such kindness, when there is any coolness between the two animals, and especially if the animal from which the other is begging is in a bad humour. Nobody could look more undisturbed, more indifferent, more uninterested, than, generally, a chimpanzee who is eating, and of whom another, with outstretched hands and pleading voice, is begging for some of his food. Even when the pleader distressfully beats the air with his arms, or throws himself whimpering on his back, the other seems not even to notice him; he chews away, and looks quite imperturbably past the pleader. That he sees him at all is evidenced only by the fact that from time to time he squeezes the food between his feet, or covers it with his arm, when the animal that is pleading gets his hand too near. But this scene may have a different ending, when the pleader is a good friend of the rich ape, and the latter happens to be in a good mood. In that case the ape will let the other slowly take a little fruit from the ground, or even seize it out of his hands. If this procedure may have several interpretations, because a passive attitude like this may create the impression that the main point here is to get rid of an importunate beggar, there are yet enough cases where the whole attitude of the giver is a picture of pleasant friendliness. He will suddenly gather some fruit together and hold it out to the other—as I have seen dozens of times—or he will take the banana which he was just going to put in his mouth, break it in half and hand one piece to the other ape, eating the rest himself. Once when Sultan, for the purposes of a test, was put in a special room and kept on a diminished diet, the following took place at the others' feeding-time : Tschego, as soon as she had got her lot of bananas, squatted in her usual feeding place, about three metres from the bars behind which Sultan was locked, and, turning her back on him, began to chew lustily and smack her lips with enjoyment. Sultan, who had had nothing, began to lament, first softly and then louder and louder, excitedly scratching his head and his breast. He stretched his arm out to the big animal, took some little stones and stalks, and threw them, in the way apes have, in the direction of what he wanted.[1] Finally he jumped up and down behind the bars, in the utmost impatience, screaming as if he were being tortured. The female chimpanzee suddenly got up, gathered together a heap of bananas, took a few steps towards the other, and handed them to him through the railings. Then she went back to her interrupted meal. Of course, the isolated ape vigorously

[1] Cf. above, pp. 88 seqq.

began his lamentation the next day as soon as Tschego began
to eat, and actually succeeded in making Tschego feed him
in this way for five days. But on the sixth day the big female
took no notice of the little male's loudest cries; she had
probably lost interest in him, for on the sixth day the cold
phase of Tschego's sexual period (which was recognizable
objectively) began, and it was not until this was over that
her former behaviour recurred.

I have not been able to obtain an altogether adequate notion
of the sexual behaviour of chimpanzees. As in other depart-
ments of their activities, so here : the sexual life of the Station
group would have developed on somewhat different lines if
at least one adult male had been among them from the first.
There appears, however, to be among chimpanzees nothing
resembling the unrestrained and all-absorbing sexual impulse
which is attributed to some species of monkeys. It is true
that the young males, six or even eight years before maturity,
as little boys, meeting the adult female, Tschego, at her
instigation became excited and went through movements
of coitus with her ; but certainly one could not say that they
were driven by an uncontrollable impulse under those circumstances. It seems to me that among these animals sexual
excitement is less specific, and less differentiated from any
other kind of excitement, than among human beings. We
may almost say that any strong emotion, and thus also any
strong external stimulus tends to react directly upon both
the colon and the genitals, but not so as to give the impression
of exaggerated and concentrated sexuality, but rather of an
extreme vehemence and interdependence of all vivid inner
processes. We may even say that this frequency of sexual
effects implies a certain trivialization of this sphere of life,
rather than an accentuation. I admit that if—from motives
of hygiene—one prevents coitus among them by segregating
the sexes, one encounters developments which would not be

likely in their natural condition, especially among the females in a state of need.   Thus it was simply owing to our prohibition that Sultan did not at once copulate with Tschego in óur presence, but in obvious collusion with Tschego and at her invitation, retired after her or led the way to a hiding-place.

The sexuality of these animals is fairly diffuse, especially in so far as there is not, either before or after puberty, any *absolute* differentiation or orientation according to sex.   Often a female before the period of maturity will execute movements of an unmistakable character towards another young female ; one will adopt the attitude and actions of the male. Later, when fully mature, during the very pronounced recurrent swelling of the genital organs, they press themselves against one another and perform mutual friction.   No doubt this was partly owing to the lack of a due number of adult males, but I can only repeat that even the strongest expressions of sexual behaviour gave a very naïve impression, and the drive remains naïve since under its normal conditions of functioning, it merges completely into the rest of the " social ", or communal life of the group.   The sexuality of the chimpanzee is as it were less *sexual* than that of the civilized human being. Often when two chimpanzees meet one another, they seem to " sketch ", or indicate, movements, which can hardly be classed definitely under either the category of joyous and cordial welcome, or that of sexual intimacy.

[The female of the species definitely menstruates, at intervals of thirty to thirty-one days, and always for a period of between three and six weeks.   During the flow her sexual instinct is absolutely quiescent, but her temper is often particularly amiable.   After the cessation of the flow, there is an access of sexual desire, accompanied by an enormous swelling around the genitalia.   At this time the animals are irritable and uncertain in temper, and suffer a good deal from the very sensitive swollen area, until it subsides.

I note, in conclusion, that Grande, who was, in many respects exceptional, always showed considerable sex indifference towards her male fellow-captives and was also " let alone " in this respect by them, although otherwise they were perfect comrades and play-fellows.]

Distinctively sexual approaches often actually appear as friendly greetings between these creatures, though there is a good deal of variety here. Thus chimpanzees *embrace* each other, with all degrees of emphasis and fervour, partly as reassurance of their social cohesion, but also as a consolation in moments of terror and anxiety, and again " just because life is so jolly ". In moments of special cordiality they fall over each other to the ground. Another very friendly form of welcome, to which I have already alluded, proceeds as follows : If one animal is on its feet and the other squatting, the former will sometimes place his hand on the groin of the latter ; or the seated ape will grasp his companion's hand, place it between his thigh and abdomen, and pat it with his own hand. It is customary among them, when greeting a friend of the same species who is in a standing position, to place their hand between his or her thighs ; a female, in the condition of local swelling and sensitiveness which follows the period, is greeted by being gently clasped from behind in the genital region and will often invite the greeting by protruding her hind quarters—again a case of the " borderland " between sociability and secondary sexual manifestations. Mutual hand-clasping is not used as a form of greeting, but appears sometimes as a spontaneous expression of joy and sympathy on special occasions. Thus, it has happened that two of the apes, as they sit opposite one another and in front of two great heaps of green stuff, chewing their food with indescribable gusto, seize each others' hands in their enthusiasm at the delicious feast. There is yet another form of greeting, which appears to have a special emotional value. An arm is extended with the hand flexed inwards and towards the ape himself, so that the back of the hand is towards the person greeted, and the fact that a *human* friend is especially often greeted in this way, seems to give this greeting a special character. When a chimpanzee approaches another of the same species with whom he is on a " difficult " footing—for instance, if they have recently been fighting—and is dubious about the possible reception of his advances, he will probably extend his hand with the palm turned *inwards* as described. I am not absolutely sure about the significance of this gesture, as it may also often be observed in circumstances of complete tranquillity. But one may perhaps guess that the flexion of the palm and the extension of the back of the hand are meant to reassure, by contrast to the grasping or hacking motion characteristic of attack. For the chimpanzee takes care as a rule that friendly overtures on his part are quite unmistakable. Thus, when he presses the fingers of his human friend between his lips in good-humoured play he draws down the flesh of his jaws over his great teeth.

It is difficult to describe the methods of inter-communication among these animals, apart from their greetings. It may be taken as positively proved that their gamut of *phonetics* is entirely " subjective ", and can only express emotions, never designate or describe objects. But they have so many phonetic elements which are also common to human languages, that their lack of articulate speech cannot be ascribed to *secondary* (glossolabial) limitations. Their gestures too, of face and body, like their expression in sound, never designate or describe objects. But their range of expression by gesture and action is very wide and varied, and, beyond all comparison, superior, not only to that of the lower apes, but also to the orang-utan's. Much is easily comprehensible to us human beings—for example, rage, terror, despair, grief, pleading, desire, and also playfulness and pleasure. But, in photographs, the expression of a slight degree of fear is often mistaken for mirth, and great fear for rage—though when a genuine manifestation of rage occurs, it is unmistakable[1]. The expression of other feelings becomes intelligible to the careful observer within a few weeks, with one exception : there are certain spells of " pure excitement " over whose exact character I have formed no definite opinion, even after six years' study[2]. But among themselves the animals understand perfectly " what is the matter ", on almost every occasion—that is evident from their communal behaviour towards the individual in question ; and we psychologists, who in theory are accustomed to derive such understanding from inferences by analogy or by reproduction of our own experienced mental states, are reduced to a theoretical helplessness and confusion which is in strong contrast to the sure and clear mutual comprehension of the animals. Even quite apart from this unsolved problem in psychology, we must consider the exact phenomenology of expression among chimpanzees as an important field for future research, as the study of the connexion between moods or emotions, and their expression, is especially easy in the case of these highly excitable creatures. Orangs, for example, have either a meagre emotional life, or are constitutionally unable to give it such vivid bodily expression.

The chimpanzee's register of emotional expression is even greater than that of average human beings, because his whole body is agitated and not merely his facial muscles. He jumps up and down both in joyful anticipation and in impatient annoyance and anger ; and in extreme despair—which develops under very slight provocation—flings himself on his

[1] Human beings in the extremity of pain open their mouths wide and twist them sideways, so that their teeth are seen, as the chimpanzee does in the extremity of fear.

[2] My predecessor at the Tenerife station, Mr. E. Teuber, pointed out to me that such spells of excitement occurred fairly often, and that we could not ascertain their exact emotional *tone* or direction.

back and rolls wildly to and fro. He also swings and waves
his arms about over his head in a fantastic manner, which may
not be unknown among non-European races, as a sign of disappointment and dejection. I have never seen anthropoids
weep, nor laugh in quite the human sense of the term. There
is a certain resemblance to our laughter in their rhythmic
gasping and grunting when they are tickled, and probably
this manifestation is, physiologically, remotely akin to *laughter*.
And, during the leisurely contemplation of any objects which
give particular pleasure (for example, little human children),
the whole face, and especially the outer corners of the mouth,
are formed into an expression that resembles our " smile "[1].
I have already mentioned the habit of scratching the head
when uncertain and in doubt[2], but to scratch the whole surface
of the body, especially the arms, the breast, the upper portion
of the thighs, and the lower abdomen, and against the direction
in which the hair grows, is expressive of a wide diversity
of emotions ; we have nothing exactly corresponding
thereto, at least among Europeans ; after all, we no
longer have fur or hairy covering long and dense enough
to stand on end so effectively as a chimpanzee's in several
emotions.

Chimpanzees understand " between themselves ", not only
the expression of *subjective moods* and emotional states, but
also of definite desires and urges, whether directed towards
another of the same species, or towards other creatures or
objects. I have mentioned the manner in which some of them
used the " language of the eyes " when in a state of sexual
excitement. A considerable proportion of all desires is
naturally expressed by slight initiation of the actions which
are desired. Thus, one chimpanzee, who wishes to be accom-

[1] Though the chimpanzee at once correctly interprets the slightest
change of human expression, whether menacing or friendly, he seems
permanently incapable of understanding merry human laughter.
[2] *Optische Untersuchungen*, etc., p. 16

panied by another, gives the latter a nudge, or pulls his hand, looking at him and making the movements of " walking " in the direction desired. One who wishes to receive bananas from another initiates the movement of snatching or grasping, accompanied by intensely pleading glances and pouts. The summoning of another animal from a considerable distance is often accompanied by a beckoning very human in character. The chimpanzee has also a way of " beckoning with the foot," by thrusting it forwards a little sideways, and scratching with it on the ground. Human beings are often invited by a gesture of what the animals want done ; thus, Rana, when she wished to be petted, stretched her hand out towards us, and at the same time clumsily stroked and patted herself, while gazing with eager pleading. Another obvious method of invitation is for an ape to assume or indicate in his own person whatever movements he would perform in the activity he wishes the other to undertake, in the same way that a dog invites us to play with him, by leaping and running, and then looking back towards us. Anthropoids behave in the same way in inciting others to play with them, to have sexual relations with them, or to join with them in that mutual inspection of the skin and hair, which is one of their most absorbing occupations ; in all cases their mimetic actions are characteristic enough to be distinctly understood by their comrades. When I grew tired of tickling Tschego's back and ribs, which always delighted her, she would still come and stand before me in that crouching position typical of a human being or anthropoid who is being tickled, and would make the half-defensive gestures belonging as well to this nice situation. But there is a sharp barrier to mutual comprehension, when one of these apes sees another executing intelligent new actions quite unusual among chimpanzees.

Every visitor to Zoological Gardens knows that monkeys of many kinds have a custom of mutual personal " inspection,"

and skin treatment[1].   But we are at present in the dark as to
the reasons for the popularity of this investigation of skin,
hair, and hind quarters, which is carried out with the greatest
eagerness and attention—and is a pleasure shared equally by
the active and passive parties in the process.   I am inclined
to rank this manifestation, together with the constant urge
towards nest-building, among those racial characteristics
which we do not explain at all by their title " instincts," this
being merely the label for a complex of equally peculiar
biological riddles.   This skin treatment is distinctively *social*,
no chimpanzee shows so much interest in his own body if he is
alone.   There seems to be quite a specialized technique here ;
the mouth of the active partner is rapidly opened and shut in a
way peculiar to this activity, but, so far as human observation
goes, in no direct relation to the skin treatment itself, for it
does not happen very frequently that a loose bit of skin is
lifted to the mouth ; such a tiny object would not require the
gnashing movement of the jaws for its mastication, and that
movement is begun quite independently.   I am not aware
whether there is anything analogous among primitive tribes—
a tendency to take great satisfaction in mutual *manipulations
of all the parts of the body*, and to spend much time in this way.
It is known, however, that many such primitives like to vary
their looks by the simple method of tearing out the hairs ; but
whether this procedure gave originally a form of *social* satisfaction, or was only intended to produce the external effect,
we are quite unable to tell.   Curiously enough, our chimpanzees in Tenerife developed a passing fashion of tearing out
handfuls of each other's hair, on the head, back, and shoulders
—not in malice or battle, but in extension of the skin treatment described above.   The ape being " plucked " remained
quite still and passive during the process.

[1] This means real skin treatment, not " lousing ", or the removal of
parasites.

As a branch of the same activity, the chimpanzee likes to pay attention to wounds or injuries received by his fellows ; but hardly urged thereto by motives of " mutual aid ". To handle such things gives him pleasure, and there is sometimes a helpful and beneficial result. Once an enormous abscess had appeared on the lower jaw of one of our chimpanzees. When it became noticeable through the extent of the inflamed surface and secretion of pus, another of the apes would not stir from the patient's side, but pressed and kneaded the injured jaw, until the pus was removed, revealing a raw, gaping wound. The animal thus treated made no objection. As apes like using diverse objects in all eager manipulations, the operator worked with a large piece of old rag in his hand. Yet, wonderful to relate, the wound—itself probably originally caused by skin treatment with filthy hands—healed rapidly and completely. Chimpanzees also like very much to remove splinters from each other's hands or feet, by the method in use among the ordinary human laity. Two finger-nails are pressed down on either side and the splinter levered upwards, to be caught and removed by the teeth. At the risk of infection, I went up to a chimpanzee on one occasion when I had run a splinter into one of my fingers and pointed it out to him. Immediately his mien and expression assumed the eager intensity proper to " skin treatment " ; he examined the wound, seized my hand and forced out the splinter by two very skilful, but somewhat painful, squeezes with his finger-nails ; he then examined my hand again, very closely, and let it fall, satisfied with his work. This example proves that a human friend is treated very much like a member of the same species ; if one tries to perform " skin treatment " for a chimpanzee, he will gladly return the compliment by inspecting one's skin and hair. And if one is on friendly and familiar terms with an ape who has been injured—say by a bite—one can easily induce the animal to extend the injured limb or surface for inspection, by making the expressive sounds which indicate sorrow and regret, both among us and among the chimpanzees.

III

The chimpanzees became more and more indolent after they had attained sexual maturity ; often they lay about in a sort of slumber nearly the whole day, and only roused themselves under the stimulus of meal-times, or some intervention from outside. Perhaps their long captivity exercised an influence in this direction, as well as biological factors. But at the beginning of their time with us, their vivacity was unimpaired, and they occupied themselves in various forms of continuous play. I have already described several such activities[1] ; here I will only give a few further details. Nueva was especially ingenious. Having once discovered that it was possible to dip up water out of the butt with her little drinking-cup, she incessantly dipped and filled the cup and then poured back the water into the butt. She hardly drank it at all, but even the drops that ran down the cup were of interest to her, and she loved to dip her hand into the water and watch the rain of drops fall from it. She also used her bread—for which she did not care very much—in this water game ; she dipped and soaked it and then sucked the water from it ; dipped and soaked again, and so forth. She was also an indefatigable "collector." She scraped together stones, pieces of wire and wood, rags and banana skins, into heaps on the ground, into her nest, or into a tin bowl, and seemed to derive the greatest satisfaction from this procedure. None of the other animals had so developed a taste for collecting and putting objects together. Three days after her arrival in Tenerife, she split a

[1] See above, Chapter III, etc.

wooden plank with her teeth and drove a wire into the gap ;
on the following day she was busy with a woollen rag which
she tied to a stick ; she was not content with simply wrapping
it round the stick, but actually achieved a sort of knot, by
looping one end of the rag through the portion wound round
the stick and pulling it taut. However humble this effort
may seem to most of us, it has a surprisingly *constructive*
character for anyone who knows the tearing, smashing, and
demolishing tendencies of the species. Other apes than
Nueva also liked to poke about with straws (or sticks) in holes
and crevices, but I never observed any of them " weaving "
and carefully plaiting straws through the wire interstices as
she did. She had a special fancy for knots ; for instance,
she thrust a strip of banana leaf through a wire mesh, laboriously drew the end back through another mesh, tied the two
ends together, and continued in the same way, either by
slipping one end of the leaf through the knot, or tying the ends
again. I often thought that she was about to begin a deliberate, though rudimentary, constructive effort, a form of manual
craftsmanship, but she could never be induced to continue
these efforts on any plan, however easy. When I prepared for
her a wooden frame with strips of leaf fastened to one side of
it for plaiting, she turned aside and devoted herself to her usual
knots ; the slightest pressure towards anything stable and
" productive " extinguished her joy and interest at once, and
she let the frame fall in sullen displeasure.

Nueva thought out these forms of play, during a long time
of isolation. Sultan, who was also much above the average
in intelligence, performed extraordinary antics with his own
limbs and body, during times of isolation. Often, as he
squatted on the ground, he would take hold of one of his own
legs with both arms, stroke it, rock it to and fro, and generally
treat it as some pleasant, but wholly exterior, object. Or he
would stretch out either one or both legs on the ground, limp

and motionless, and shuffle along on his powerful hands.[1] There are several such " fancy " methods of locomotion among chimpanzees. In their usual walk, the hand is placed on the ground in such a way as to touch it only with the fingers, which are bent inwards. Suddenly, however, in play, one of them will walk around, touching the earth with the whole palm, and remain in that posture for a while. Upright walking (without brachial support) takes place when the hands are full, when the ground is wet and cold, or when the animals are excited in various ways. But in special individual chimpanzees, whose peculiar build is well adapted to the upright posture, it may become a "fashion", a form of play that persists for days. I recollect seeing somersaults turned by both orangs and chimpanzees. Chimpanzees also sometimes lie down at full length on the ground, and revolve with vertiginous speed, round their own length, to a distance of many metres. When in so doing they wrap themselves in a blanket, or creep into a sack beforehand, the impression they produce is comic in the extreme, and their fellow-apes play all sorts of pranks with the rolling bundle. When they play in groups of two, three, or more, these games assume still more different forms ; one will lie limp and motionless, a friend seize his arm or foot, and drag him along like a corpse or inanimate mass. Or a little ape will leap onto the shoulders of a larger one, and make him carry him, and then slide gradually forwards to his neck, drop on his hands, and march solemnly along with him like a new six-legged monstrosity.

Rothmann and Teuber have described a curious habit of these animals ; they sometimes tear along, as if possessed, by the walls of their sleeping-dens, and kick them until the excitement subsides. My colleagues consider this to be a sort of

---

[1] This was not the trivial variant of the chimpanzee's mode of locomotion, in which the hands support the body, and the flexed lower limbs are swung forwards between the arms.

sex dance. To me it has never looked like this, but rather as the explosive culmination of one of those strange fits of excitement, whose character and origin are still riddles to us. And I am the less inclined to speak of *dance rudiment* here, as there are other activities of these apes, quite unlike their frantic " fits " and easy to understand, which one might far more easily call " primitive stages of dancing "[1]. One lovely fresh morning Tschego and Grande were playing together on a box. Presently Grande rose upright, and with bristling hair, in her characteristic, pompous, and would-be-terrible manner, began to stamp first one foot and then the other, till the box shook. Meanwhile, Tschego slipped from the box, rose upright, and slowly revolved round her own axis in front of Grande, springing clumsily and heavily—but springing—from one foot to the other. They appeared to incite each other to these strange antics and to be in the best of tempers. I have frequent notes of such behaviour. Any game of two together was apt to turn into this " spinning-top " play, which appeared to express a climax of friendly and amicable *joie de vivre*. The resemblance to a human dance became truly striking when the rotations were rapid, or when Tschego, for instance, stretched her arms out horizontally as she spun round. Tschego and Chica—whose favourite " fashion " during 1916 was this " spinning "—sometimes combined a forward movement with the rotations, and so they revolved slowly round their own axes and along the playground.

The whole *group* of chimpanzees sometimes combined in more elaborate *motion-patterns*. For instance, two would wrestle and tumble about playing near some post ; soon their movements would become more regular and tend to describe a circle round the post as a centre. One after another, the rest of the group approach, join the two, and finally they

[1] Rothmann and Teuber have also observed and described these activities.

march in an orderly fashion and in single file round and round the post. The character of their movements changes ; they no longer walk, they trot, and as a rule with special emphasis on one foot, while the other steps lightly ; thus a rough approximate rhythm develops, and they tend to " keep time " with one another. They wag their heads in time to the steps of their " dance " and appear full of eager enjoyment of their primitive game. Variations are invented time and again ; now and then an ape went backwards, snapping drolly at the one behind him ; often the circular common movement would be varied by individuals spinning round their own axis at the same time ; and once, as the whole group were joyously trotting round a box, little Konsul stepped to one side outside the circle, drew himself up to his full height, swung his arms to and fro in time to the trotting, and each time that fat Tschego passed him, caught her a sounding smack behind. A trusted human friend is allowed to share in these games with pleasure, as well as in other diversions, and sometimes I only needed to stamp rhythmically, as described, round and round a post, for a couple of black figures to form my train. If I had enough of it and left them, the game generally came to an abrupt end. The animals squatted down with an air of disappointment, like children who " won't play any more," when their big brother turns away.

[I first took part in this game after it had taken place without me hundreds of times. It seems to me extraordinary that there should arise quite spontaneously, among chimpanzees, anything that so strongly suggests the primitive dancing of some primitive tribes.]

Even in the course of such simple activities, incredible as it may seem, decided individual differences were made manifest. It is obviously much more interesting to trot round two posts or boxes, standing at some distance from each other, than to trace a mere circle round one such centre. But now the " single file " spreads and straggles, there are gaps between the animals in the large ellipse, and one must be able, subjectively, as it were, to complete the form of the ellipse, if one is not to fall out when sharing in this game. This was too much to expect of the eager, but always foolish Rana, who, whenever the circle extended itself to the more difficult form, would get confused, fall out into the interior of the ellipse, and suddenly, to her own surprise, knock against a fellow-ape, who had kept on the right track, and circumnavigated the second post.

[I have already mentioned, among other forms of play, the chimpanzee's passion for adorning himself, especially with dangling strings, rags, or blades of grass. It is in accordance with its play-character that this decorating should often occur when the apes are " dancing ", or also, that the decoration should produce a strutting to and fro, and all those primitive rhythmic activities of the group. Whereas I formerly stated that up till now there had been—with only one exception—observation and record of *playful* self-adornment only, I must now admit, on the basis of longer experience, that the protective covering of the chimpanzee is not confined to the human blanket in which he wraps himself up from the cold. At the first drops of a cold shower of rain, we may see the animal look upwards, and at once seize some bits of grass, leaves, etc., lay them on the back of his neck and pat them firmly down with his hands, as if to stick them fast. They afford, of course, no real protection at all, but as in many other situations[1], so here ; the chimpanzee does something orientated in the *direction* of his need, and under the urge of that strongly felt need, so that the action is more of an emotional expression.]

[1] Cf. above, p. 92. Such observations may also be found in Reichenow : *Naturwissenschaften*, IX, p. 73 seqq., 1921.

IV

It has been recounted that some monkeys, dogs, cats, and even birds, when faced by their own reflections in a mirror, react—even if only momentarily—as though a real individual of the same species stood before them. When we gave the chimpanzees a hand-mirror for the first time, they looked into it and at once became intensely interested. Each one wanted to look, and tore the wonderful object out of the other's hand : and I was only able to observe methods of proceeding with both mirror and the picture behind it, when eventually Rana captured the hand-glass and escaped with it to a remote corner of the roof. She gazed long and intently into the mirror, looked up and then down, put it to her face and licked it once, stared into it again, and suddenly her free hand rose and grasped—as though at a body behind the mirror. But as she grasped emptiness she dropped the mirror sideways in her astonishment. Then she lifted it again, stared fixedly at the other ape, and again was misled into grasping into empty space. She became impatient and struck out violently behind the mirror ; finding this, too, in vain, she " lay in wait ", after the manner of chimpanzees when they watch (with the most aloof and harmless expression in the world) whether anyone outside their cage will touch the bars in a heedless moment. She held the mirror still in one hand, drew back the other arm as far as possible behind her back, gazed with an air of indifference at the other animal, then suddenly made a pounce at him with her free hand. However, both she and the rest soon became used to this side of the affair, and concentrated all their interest on the image ; this interest did not decrease —as in the case of other species enumerated above—but remained so strong that the playing with reflecting surfaces became one of the most popular and permanent of their

" fashions ". Soon they dispensed with the human implement; having once had their attention drawn to it, they mirrored themselves in anything at all available for the purpose : in bright pieces of tin, in polished potsherds, in tiny glass splinters, for which their hands provided the background, and, above all, in pools of water. I have often observed Tschego for long at a time sunk in contemplation of her own reflection in a pool. She played with it : bent far over it and drew back slowly, shook her head backwards and forwards, and made all kinds of grimaces over and over again. Finally, she dipped her great hand into the puddle, shaking and wagging her head, and let the water trickle back onto the picture in the water. As the animals were constantly looking at themselves, using even tiny surfaces, which we humans would never have thought of using for this purpose, they developed a pleasant and inter-esting extension of their play. They slowly turned the reflecting surface, or moved their heads to one side, so that they could no longer see themselves, but continued to look into it, examining the images of one object in the room after another, with unabated interest, and it could constantly be observed that as they turned the " mirror " they glanced quickly from time to time towards the real, and, of course, familiar, everyday objects that had just appeared behind it. This is, of course, not really strange, as the image of well-known inanimate things must have been just as dis-tinct and clearly recognizable as their own. The following example will prove how absorbing was the phenomenon of reflection to them. Their sleeping-dens had barred windows, without any glass panes in them. The floor was of smooth cement. When they retired to rest at night, their urine was often deposited on the flat floor, where it formed shallow puddles. As soon as this occurred, one or other of the anthro-poids could be observed bending sideways, with eyes fixed on the liquid, and moving his head slowly to and fro in order to catch the reflection of objects from outside the window. Other animals soon lose interest in the reflections when non-optical control proves their "unreality." What strange beings are the chimpanzees, to be permanently attracted by the contemplation of such phenomena, which can bring them not the least tangible or "practical" benefit.

The behaviour of chimpanzees towards other animals varies according to the particular appearance or manner of the creature in question. Dogs, who rushed up at them outside the bars, leaping and barking excitedly, were promptly teased by kicks and jumps at the bars, by stone-throwing and stick-thrusting : the anthropoids gave no sign whatever of fear towards them. They were somewhat more reserved and cautious in their reception of a cat, who one day suddenly made her appearance in their playground. One or two slowly approached within a few steps and made half-playful, half-menacing gestures, standing upright and rocking to and fro from one foot to the other. But when the cat, "having had enough of it ", arched her back and spat at them, the most daring apes retreated very quickly.

Even a practically defenceless creature can generally protect itself against these anthropoids, if only its appearance arouses sudden terror. When they were behind their bars every hen who ventured near them was treated as a toy, and often brutally enough, as were also such fowls as ventured into the play-ground. But one day Sultan wandered into a poultry yard, in which a proud mother was taking her tiny chickens for a walk. He approached them, but she flew at him with all her feathers ruffled, in the well-known hen attitude of defensive wrath, and in a second he had leapt the partition and was gone.

[In general we may say that not only what has been experienced, or may be recognized as really dangerous, inspires fear in these animals, but also anything which has the phenomenological character of aggressiveness and "awfulness"—

especially when there is the added factor of the surprising
and unknown. The same axiom holds good among themselves. When a small weak chimpanzee has become blind
and reckless through intensity of anger, he can drive a much
larger and more powerful comrade in headlong flight before
him.]

Large and uncommon animals caused a perfect panic if
they merely came into the neighbourhood of the apes. When
once, two of the enormous oxen of Tenerife appeared, drawing
the primitive plough of the island up and down outside in the
farm, the whole group of chimpanzees tore like creatures
possessed, first one way, then another, each time as far as
possible in the opposite direction from the monster; tremb-
lingly they hid their faces grown livid with fear. And indeed
no purgative could have acted more drastically than did the
spectacle of these ruminants ! Some camels only needed to pass
by once, to make any experiments with the chimpanzees quite
impossible for a long time : an anxious and absorbed attention
was turned exclusively in the direction from which the camels'
bells were heard tinkling for a while after they themselves
had passed out of sight.

Now there is one concept of the " biological," and the
" nearness to life," according to which it should make a great
difference in the behaviour of animals, whether an object is
familiar and at least approximately similar to parts of their
natural surroundings, or is such as none of their species can
ever have had anything to do with before. In almost comic
contradiction to this is the chimpanzees' reaction to the very
crudest reproductions of any other kind of animal. I tested
them with some most primitive stuffed toys, on wooden frames,
fastened on to a stand, and padded with straw sewn inside
cloth covers, with black buttons for eyes. They were about
forty centimetres in height, and could perhaps be taken for
caricatures of oxen and asses, though most drolly unnatural.

It was totally impossible to get Sultan, who at that time could be led by the hand outside, near these small objects, which had so little real resemblance to any kind of animal. He went into paroxysms of terror, or threatened recklessly to bite my fingers, when I, whose dangerous possibilities were well-known to him, tried to draw him towards the toy, as he struggled and strained backwards. One day I entered their room with one of these toys under my arm. Their reaction-times can be very short ; in a moment a black cluster, consisting of the whole group of chimpanzees, hung suspended from the farthest corner of the wire-roofing, each individual trying to thrust the others aside and bury his deep head in among them. On another occasion, in the morning, I placed one of these stuffed donkeys in the apes' stockade, and laid the banana bunch (the breakfast) under him on the wooden board which supported his legs. The apes crept together into a corner and only occasionally did one of them venture a terrified glance at the dreadful Being. About half-an-hour elapsed before big Tschego, after many bold resolves to approach and an equal number of retreats halfway, finally ventured to snatch *one* banana from under the donkey's tail, and to tear off with it at top speed. When the stuffed animal is smaller, and of more reassuring aspect than those particular toys (whose motionless, protruding, black button eyes probably inspire peculiar dread) the fright displayed is less ; but quite recently, in Berlin, a light-coloured, friendly-looking little toy-horse, of very modest dimensions, was treated with respectful awe, and even after Grande had cautiously pushed it over with her finger-tips, the others did not approach the uncanny creature.

It is too facile an explanation of these reactions to assume that everything new and unknown appears terrible to these creatures. Any geometrical figure of wood found standing or lying about, though it represents something quite new to them, rouses no such convulsions of terror, even though in the first moment it is rather cautiously examined. New things are not necessarily frightful to a chimpanzee, any more than to a human child ; certain impressive qualities are requisite to produce this special effect. But, as the examples cited above prove, any marked resemblance to the living foes of their species does not seem at all essential, and it almost seems as though the immediate impression of something exceptionally frightful could be conveyed in an even higher degree by *constructing* something frightful, than by any existing animal (with the possible exception of snakes). For us human beings as well, many ghost-forms and spectres, with which no terrible *experience* can be individually connected, are much more uncanny than certain very substantial dangers, which we may easily have encountered in daily life. This particular section of emotional psychology is as yet very obscure. It has probably a wider significance than is yet manifest to our superficially empiristic theory of perception, especially as it is impossible to deduce those effects from the experiences of ancestors. It is not an admissible hypothesis that certain shapes and outlines of things have in themselves the quality of weirdness and frightfulness, not in consequence of earlier experience, and not because any special mechanism in us enables them to produce it, but because, granted our general psychophysical nature, some shapes necessarily have the character of the terrible, exactly as others the character of grace, or clumsiness, or energy, or decidedness. I may mention that once, as Sultan squatted in front of a box, into which I had just put a couple of bananas—under his very eyes and opening the lid to its fullest extent—he did not at first dare to put his hand into a hole in the side of the box, which looked black against the high brightness around it, but every time pulled it back just as it reached the hole.

One day, as I approached the stockade, I suddenly pulled over my head and face a cardboard copy of the mask of a

Cingalese plague demon (certainly an appalling object),[1] and instantly every chimpanzee, except Grande, had disappeared. They rushed as if possessed, into one of the cages, and as I came still nearer, the courageous Grande also disappeared. Had I held a piece of blank cardboard before my features, it might also have seemed rather unpleasant, but, guessing from their behaviour, when we change our clothing, etc., it could never have caused such a panic.

[I received the following further proof that we do not see these things rightly with our concept of the " biologically adequate ", or at least that we do not apply this concept correctly, if we assume the experiences of the individual or the species to be decisive for the development of such reactions. A dog trotting along the street suddenly came across that miniature stuffed donkey. He sprang back a distance of a few yards, sprang forward barking, again retreated undecidedly and began to circle around at a greater or less distance, alternately barking and growling the whole time ; so he voiced his suspicions until he had plucked up sufficient courage to sniff at the hind quarters of the " animal," whereupon he at once became quite indifferent. This final reaction is more in line with our ideas about the importance of " nearness to life " in animal reactions ; but how was it that the dog regarded a quite unnatural toy as dangerous on first sight ? It is perhaps even more extraordinary that a real donkey, which often passed by the little toy at some distance, never failed to show all the signs of interest and excitement—unmistakably directed towards the toy—which are peculiar to his species when its members meet one another.

Again, the following incident shows how closely some of the perceptions of the higher animal species must resemble ours, though in this instance one might at first tend to regard human perception as the product of very complicated " higher pro-

[1] Cf. Schurtz, *Urgeschichte der Kultur*, Plate on p. 117, No. I.

cesses ". I was riding down a mountain path on a cold moon-light night in winter, when a faint mist made everything seem important and fantastic in outline. My guide was walking behind me, and his very lively little dog trotted wearily about one hundred metres in front. A little ahead the path dipped into a dark cleft in the hills ; facing us in the moonlight several pine stumps stood out upon the opposite slope. As I rode on I perceived, to my surprise, that an old man was squatting in an attitude eloquent of misery and dejection on the top of one of these stumps, alone and motionless in the cold night among the mountain solitudes ; a pitiful and, at the same time, a somewhat uncanny spectacle. As our pathway led close by the old man's perch, I waited in silence till we should reach him. The guide seemed to have noticed nothing—at any rate, he, too, was silent. When the dog had got near to the crouching form, he sprang with an excited bark towards him, and circled round for some time, still barking. Then he appeared to regain his composure and went on his way. Meanwhile we had come up, and in a few rapid processes which I cannot describe, the old man had resolved himself into a group of short branches and shoots of one of the pine stumps. I then asked the guide why the dog had barked—they were the first words uttered. " Oh, he probably thought that there was an old man," was the answer. " I thought so, too, at first." For two human beings, at the same time but quite independently, to have received the same optical delusion the complex of optical stimuli must have had a very compulsory effect at some distance ; and now the dog, who trotted heedlessly past dozens of pine stumps—some of them with strongly developed twigs and shoots—was much excited under the same con-ditions. Even if we admit that he may not have received exactly the same impression as we did, the occurrence remains curious enough.

Both images in a mirror and children's animal-toys do,

after all, possess considerable likeness to their originals, in so far as they both display *colour* and have three dimensions. It was an obvious step farther along the same lines, to test the effect of flat and colourless reproductions, using photographs for the purpose. I showed the chimpanzees, individually and separately, some photographs of themselves or members of the same species, fairly good likenesses taken with a good camera in the dimensions of eight by ten and a half centimetres, and representing the animals from four to eight centimetres high. They were examined with great attention, and the glance and focussing of the apes' eyes were not those with which a cursory look is cast at any piece of paper, but those with which a small object is examined in detail ; it is well known that the two kinds of looking are typically unlike. Tschego at once took her own portrait, gazed at it carefully, passed her hand over it, turned it over for a moment so that the blank obverse was visible, put it in the fold of her lap and carried it away. Grande, to whom I showed a photograph through the bars of her cage, was equally attentive, and tried, by twisting her head sideways, to catch a glimpse of the other side, again and again. The others behaved similarly. But when it was Sultan's turn and I showed him his own likeness, he examined it keenly for a while, and then suddenly raised his arm and stretched out his hand towards the picture, in the specific gesture of friendly greeting I have already described— palm inward. He did this repeatedly when I again showed him the photographs, and his intention was unmistakable, for, as soon as I displayed the blank obverse of the photograph, Sultan simply tried to grasp it ; when I turned the picture towards him again, he repeated his gesture of greeting. I may explicitly state that Sultan had never before thus greeted an inanimate object—but always only a human being or another animal, and that I was holding the picture sideways at full arm's length, so that if he had been addressing his gesture of

welcome to *me*, he would have had to move his arm in quite
another direction ; but he did not look at me, his gaze was
fixed on the photograph.

In order to obtain more detailed results, I prepared two
photographs of the wooden banana-basket from which the
apes were fed every day ; in one of these, the box was crammed
full of this fruit, in the other it was empty and slightly at an
angle, so that the empty interior was visible.   (See Plate VIII.)
The photographs, unfortunately, did not develop very
clearly, though this does not show in the reproduction, owing
to the light background ; in the original the background
was dark.   The size of the originals was again eight by ten
and a half centimetres.   I made use of them in the following
manner : They were attached to the front of two little wooden
boxes—(in the same manner as we had used gray papers of
different brightness in earlier experiments[1])—and the boxes,
both filled with fruit, were placed in front of Sultan for his
free choice.   In ten successive tests in which no attempt was
made to influence his choice, and the positions were changed
in the usual irregular way, he every time chose the box which
bore on its front the picture of the bananas—probably because
of its greater attractiveness.   Nevertheless, his *manner* was
not very decisive, and two days later, during the same experiment, he paid so little attention, that he chose the wrong box
almost as often as the right.   As he got the food each time,
whether he chose " rightly " or not, the first faults had an
unfavourable effect, and as even the " empty " photograph
was inviting—for did it not display that important and well-
known food-basket—this result is not surprising, even if the
ape could recognize the difference between the little photographs.   I passed on to " learning " tests, counting the photo-

[1] Cf. the second and fourth reports from the Anthropoid Station,
Optische Untersuchungen und Nachweis einfacher Struckturfunktionen,
etc.

PLATE VIII. WOODEN BANANA-BASKET, FULL AND EMPTY

graph of the full basket always as " right ". Sultan rapidly
succeeded in choosing correctly in about ninety out of every
hundred occasions, though any failure of concentrated attention, when the differences were so slight and the photographs
so small, would make mistakes more probable. What did
Sultan go by? He must have acted either according to the
difference of the *objects* recognizably depicted in the photographs (i.e. the box either filled with bananas, or empty), or
else by some other differences perceived by him on the two
surfaces, as were differences of form or other optical characters[1]. In order to be sure on this point, I had two further
photographs taken which, from the point of view of forms
were quite different from the first pair.

(See Plate IX.)

One of these represented a large cluster of bananas, the
other depicted a large stone. Cluster and stone had something of the same rough outline, and the background was
the same in both cases. The likenesses were much clearer[2],
and better than in the first two photographs. In the tests
I tried single experiments with these new photographs, inter-
spersed among and between learning-tests with the old ones,
and after every decision, whether " right " or " wrong ", I fed
Sultan with bananas. His " critical " choices were thus free
from external influence. It was proved that Sultan made
more correct choices from the two fresh photographs, which
were better likenesses, and presented a stronger contrast
between the two objects represented, than the old pictures ;
it is therefore extremely probable that the determining factor
here was his recognition of the bananas as such.

In the tests with the already familiar photographs the total
proportion of " right " to " wrong " results remained as before ;

[1] Any other criteria were, of course, excluded, by the very careful
conditions of the test itself.

[2] This is not so obvious in the reproductions here, which are all about
equally distinct.

but now those choices which immediately succeeded the new set (banana-bunch and stone) showed a much higher proportion of mistakes than the others. This is explicable by the fact that the superiority of the new pictures greatly facilitated Sultan's choice, and caused an automatic relaxation of attention, so that the tests with the old photographs immediately following, found him indisposed to a close scrutiny.)

[I carried out *learning* tests on Grande, with the old photographs, and soon succeeded in getting her proficient in choosing the filled banana box, but could not make this accomplishment quite reliable and permanent, although at one time her mistakes were only five per cent. of the total. After several periods with astonishingly bad results, I terminated this set of experiments, considering that they were made independable by some factor.]

Chica learnt quite well to choose the photograph of the full banana box. But her decisions were not always certain. In the most favourable conditions she carried out several series of " choices " without any errors ; but any little interruption distracted her attention and caused a considerable number of mistakes, with these indistinct photographs. After much practice, a series of a hundred successive tests gave fifteen mistakes, and there was no prospect of obtaining better results. I therefore made the following examination of Chica's method of procedure : She had to make fifteen further selections with the first pair of photographs, and made six mistakes, showing very little interest. During a short pause, the old photographs were replaced by the very distinct and dissimilar new pair (representing the stone and the bunch of bananas), which Chica did not yet know. She did not see this change, which was performed behind a screen ; and even when the screen was removed, she did not immediately notice the change. But when she again glanced at the boxes, she *started in surprise*, remained for a few seconds motionless and staring

PLATE IX. BANANAS AND STONE

fixedly at the picture of the banana-cluster ; then, without turning her eyes away, she reached for the stick, which was used as a pointer in these tests, approached with the same intent stare, and knocked the right box hard with the stick. In the next test she showed a similar behaviour, so it was evident that there was no difficulty in her choice now. As it went so well, I tested Chica thirty-two times in succession. She made *one* mistake ; the tenth time she thrust the stick forward too hastily, but at once corrected herself when she had looked closely at the boxes.

I made no further tests, as I consider it quite obvious that results are determined simply by the technical accuracy of the photographs and the difference of the objects they repro-sent. Anyone who may take the trouble to experiment on other chimpanzees in the same way, will be able to demonstrate effectively and exactly, by means of larger and clearer repre-ductions, that the animals recognize and differentiate between such photographs. As a further variation—to meet possible objections—I would suggest, in the crucial experiments, the use of pictures of another food—say the very popular oranges or thistles—if bananas were used in the preliminary tests.

INDEX

Abdomen, objects held between it and thigh, 95

Abhand. d. Kgl. Preuss. Akad. d. Wissenschaft., 7 *n.*, 49 *n.*, 91 *n.*, 129 *n.*, 187 *n.*, 219 *n.*, 274 *n.*, 299 *n.*

Acht Vorlesungen über theoretische Physik (Planck), 209 *n.*

Adornment, love of, by chimpanzees, 91 *seqq.*, 316

Ants, eaten by chimpanzees, 77

Archives neerland. de physiol. de l'homme et des animaux, 279 *n*

Association theory, 218

Attacks, chimpanzees' alliance against, 287

Bamboo rods, fitted together, 125 *seqq.*

Basket, experiments with, 7, 18; on rod, 250

Behaviour, intelligent and unintelligent, 2, 102, 265

Bergson, H., 211

Bericht über den 5 Kongress f. exper. Psychol. (Pfungst), 222 *n.*

Boltzmann, H., 209

Boxes used for reaching objectives, 39 *seqq.*, 135 *seqq.*

Buytendyk (*Arch. neerland. de physiol. de l'homme et des animaux*), 279 *n.*

Cats, behaviour towards chimpanzees, 319

Chance, 185 *seqq.*

Chica, 5, 145; acrobatic feats, 12; with hanging basket, 18; timidity of, 19 *n.*; swings on rope to attain objective, 58; removes obstacles, 61; adept with jumping pole, 70; throws sticks, 86; carries stones, 91; uncoils rope to reach objective, 113, 115; fails to hang rope

on hook, 118; fails to empty obstacle-box, 120; uses two sticks as implement, 123; learns use of boxes, 136; tries to make box stick to wall, 157; cannot learn to use double stick, 171; uses auxiliary, 176 *seqq.*; moves string laterally to reach objective, 199, 201; with "roundabout-way board," 235, 242; with ring on nail, 246; pushes objective, towards bars of cage, 256; coprophagous habits, 298; friendly with Tercera, 299; dances, 314; tests with photographs, 328

Children, experiments with, 14, 17, 150, 243, 279; chimpanzees' behaviour towards, 295

Chimpanzees, personality among, 5; signs of anger, 6; get round obstructions, 12; highly developed vision, 36; handle objects, 67; *joie de vivre*, 69; destructiveness, 75; eating habits, 76; liking for ants, 77; behaviour towards small animals, 79; smear themselves with excrement, 80; encounter electricity, 82; quarrels of, 83; love of teasing, 83; play with and stab fowls, 84; emotions of, 88; make nests, 90; adorn their bodies, 91; sexual organs, 95; sex-cleanliness, 96; importance of optical conditions with, 108, 124, 130; conceptions of parts and wholes, 109; do not grasp complicated forms, 115; difficulties in building, 140 *seqq.*; faced with static problems, 148, 161, 163 *seqq.*; uses for

human implements, 164 *seqq.*; building in common, 166 *seqq.*; collaboration among, 169; survey the situation, 190; make no constructive solutions, 191; kinds of errors they commit, 194; narrow limits of intelligence, 204; in natural surroundings, 213; imitativeness, 222; intelligent behaviour of, 265; essential differences from man, 267; faculties of choice, 274; memory, 277; small sense of smell, 282 *n.*; dislike of solitude, 282; behaviour when sick, 284; when punished, 285; alliance against attackers, 287; absence of morphological marks in, 289 *n.*; behaviour towards new-comers, 289; to human beings 291 *seqq.*, 296 *seqq.*; mode of greeting, 294; jealousy, 296; obstinacy, *ibid.*; kindness to one another, 300; sexual behaviour, 302; menstruation, 303, 81 *n.*; intercommunication, 305 *seqq.*; emotional expressions, 306; skin treatment, 309; remove splinters, 310; indolence, 311; mode of locomotion, 313; games, *ibid.*; dancing, 314; inspect mirror, 317; fear of toys, 320; of masks, 322
"Choice boxes," selection by chimpanzees, 274
Collaboration among chimpanzees, 169
Confinement, effect of, 68
Coprophagy, 6, 80, 96, 298
Current and Magnet (Oersted), 194

Dance, rudiments of, 314
Delayed Reaction in Animals and Children, 279 *n.*
Digging, with sticks, 77
Dogs, experiments with, 13, 17, 21 *seqq.*, 27; behaviour towards chimpanzees, 319
Door, experiments with, 54
Draping, love of, 92

Ehrenfels, von, 225
"Einsicht," translation of, 219 *n.*
Electricity, chimpanzees and, 82

Emotions, of chimpanzees, 88, 306
Errors, good and bad, committed, 194, 217
Excitement, evinced by noise, 94
Excrement, chimpanzees smear themselves with and eat, 6, 80, 298
Experience, value of, 204
Experimentelle studien über das Sehen von Bewegung, 111 *n.*
Experiments: basket of fruit suspended, 7; dog and hen in a blind alley, 13; fruit in swinging basket, 18; invisible objective, 20; dog with invisible objective, 21; distant objective drawn by string, 26; basket drawn up by cord, 27; choice of strings required to draw objective, 27; objectives to be reached by using a stick, 31; objective at height from ground, to be reached by climbing on box, 39; same, but involving use of stick, 45; same, with two boxes and table, 47; ladder necessary to reach objective, *ibid.*; door to be selected and climbed upon to reach objective, 53; rope to be climbed to reach fruit, 57; obstacles placed in way of objectives, 59; objective in cage which needs to be tipped up 64; use of jumping pole, 70; objective out of reach, only to be attained by tearing off a branch and using it as a stick, 103; objective to be reached by uncoiling a piece of wire, 111; by unwinding a piece of rope, 112; by placing rope over a hook, 117; by removing stones to allow of a box being shifted, 119; by unloading a ladder, 120; small bamboo rod to be fitted into larger, 125; piece of wood to be fitted into bamboo tube, 131; building with boxes, 135; building on a heap of stones, 154; ladder-climbing, 160; use of auxiliary stick to reach objective, 174; with auxiliary tick hanging from roof, 177;

with auxiliary only reached by box which has to be unweighted, 183 ; objective only to be reached by carrying attached rope laterally, 199 ; pushing fruit past wire-netting obstacle, 226 ; with " round-about-way board," 229 ; with stick which has to be removed from nail before use, 245 ; basket containing objective hung on a nail, 250 ; objective in cage where it has to be pushed forward to bars, 252 ; use of a crooked stick, 258 ; of a T-shaped stick, 259 ; ladder lying crosswise to be brought through bars, 260 ; board to be inserted slantingly through a rectangular hole, 261 ; buried fruit to be dug up, 279 ; with mirror, 317 ; with a mask, 322 ; with photographs, 325

Fowls, teased and stabbed by chimpanzees, 84
Friendship, among chimpanzees, 299
Frogs, experiments with, 18

Games, love of, 313
Geometrical circumstances, importance of, 15, 241
Gesetze des geordneten Denkverlaufs (O. Selz), 190 n.
" Gestalt " theory, 225 seqq., 267
Gestures, use of, 308
Grande, 4, 58, 61, 70, 113 ; with hanging basket, 18 ; jumps on shoulders to reach objective, 50 ; menacing attitudes of, 83 ; love of teasing, 84 ; makes a noise when excited, 94 ; tears off branch to make implement, 104 ; raises weighted ladder, 121 ; fails to lighten obstacle-box, ibid. ; learns use of boxes, 136 seqq. ; difficulties in building, 142, 145 seqq. ; tries to make box stick to wall, 157 ; succeeds with ladder, 160 ; a good builder, 167 ; uses auxiliary stick, 175 ; her stupidities, 197 ; moves string

laterally to reach food, 199 ; with "roundabout-way board" 236, 242 ; with ring on nail. 246, 248 ; pushes basket off rod, 251 ; friendly with Tschego, 290 ; dances, 314 ; with photograph, 325 ; tests with, 328
Greeting, chimpanzee methods of, 304
Group connexions, 283

Habit, bad effects arising from, 194 seqq.
Hagenbeck, Zoological Park of, 172 n.
Handclasping, by chimpanzees, 304
Hartmann, E. von, 211
Hens, experiments with, 14
Hobhouse (Mind in Evolution), 22 n., 30 n.

Imitation, 185 seqq.
Improvement of tools attempted, 123 seqq.
Indolence, of chimpanzees, 311
Insight, 219
Instinct, blindness of, 207
Intelligence, human and animal, 3, 102
Inter-communication, methods of, 305
Invisible objectives, experiments with, 20 seqq.

Jealousy of chimpanzees, 296
Journal of Comp. Neurology and Psychology, 222 n.
Jumping with poles, 69 seqq.

Koko, 6, 26, 80, 130, 213, 215 ; with hanging basket, 19 ; uses stick to reach objective, 33, 35, 38 ; climbs on box, 41 seqq. ; builds a nest, 91 ; tries to make implement from shoe-scraper, 103 ; makes an implement, 108 ; uses auxiliary stick, 179 seqq. ; stupidities of, 197
Konsul, 5, 26, 61, 73, 139 ; jumps on others to reach objective, 49 ; his destructiveness, 171 ; illness, 285 ; friendship with Rana, 299

Ladder, for reaching objective, 47, 51 ; weighted, 121, 159 ; behind bars, 260
Lap, chimpanzees' use of, 96
Lizard, monkeys' dislike of, 80

Mask, chimpanzees' fear of a, 322
Menstruation, 81 *n.*, 303
Mental Life of Monkeys and Apes (R. M. Yerkes), 269
Mikrokosmos (H. Lotze), 94 *n.*
Mind in Evolution (Hobhouse), 22 *n.*, 30 *n.*
Mirror, chimpanzees' treatment of, 317
Motion-patterns, 314

Naturwissenschaften (Reichenow), 93 *n.*, 316 *n.*
Nests, made by chimpanzees, 90
Noise, a sign of excitement, 94
Nueva, 5, 26, 139, 213, 215 ; uses stick to reach objective, 32 ; to poke fire, 79 ; uses auxiliary stick, 175 ; with "roundabout-way board," 229, 242 ; plays with water, 311 ; collects things, *ibid.*

Obstacles, removal of, as a test, 59
Obstinacy, of chimpanzees, 296
Oerstedt (*Current and Magnet*), 194
Oetjen, F., 163 *n.*
Optical conditions, importance of, 108, 124, 130
Ornaments, love of, 91

Painting, by chimpanzees, 96
Pfungst (*Bericht über den 5 Kongress f. exper. Psychol.*), 222 *n.*
Photographs, chimpanzees with, 325 *seqq.*
Physischen gestalten in Ruhe und im stationdren Zustand, 212 *n.*
Planck (*Acht Vorlesungen über theoretische Physik*), 209 *n.*
Psychischen Fahigkeiten der Amiesen (Wasmann), 190 *n.*
Psychologische Forschung, 271 *n.*
Punishment, how received, 285

Rana, 5, 50, 61, 118, 145 ; with hanging basket, 19 ; uses sticks, 31 ; climbs on door to reach objective, 55 ; swings on rope, 59 ; fails to remove obstacle, 60 ; tries to overturn cage, 64 ; thumps with pole, 69, 72 ; her stupidity, 73, 197 ; feeds fowls, 84 ; uncoils rope to reach objective, 114, 116 ; fails to empty obstacle-box, 121 ; puts sticks end to end, 124 ; fails with auxiliary stick, 176 ; previous experiences, 205 ; moves string laterally to reach objective, 200 ; with "roundabout-way board," 239, 242 ; with ring on nail, 246 *seqq. ;* behaviour to Konsul when sick, 285 ; towards newcomers, 90 ; friendship with Konsul, 299 ; studies a mirror, 317
Reichenow (*Naturwissenschaften*), 93 *n.*, 316 *n.*
Rhythm, marching in, 92
Rothmann, 299 *n.*, 313
Roundabout methods, experiments in, 11 *seqq.*
"Roundabout-way Board," experiments with, 229 *seqq.*

Schurtz (*Urgesichte der Kultur*), 323 *n.*
Scratching, an expression of emotions, 307
Selz, O., 190 *n.*
Sequences of action, 187 *seqq.*
Sexual behaviour of chimpanzees, 302 ; organs, 95 ; swelling of organs in females, 303
Shoe-scraper, used as implement, 103
Sickness, chimpanzees' attitude towards, 284 *seqq.*
Skin, inspection of, 308
Smell, sense of, 282 *n.*
Sokolowsky, A., 172 *n.*
Splinters, removal of, 310
Stern, W., 163 *n.*
Sticks, intelligent use of, 31 *seqq.*, 73 *seqq.*, 123, 174 ; as a weapon, 82 ; throwing, 85 ; crooked and T-shaped, 258 *seqq.*
Stone-throwing, 86, 101
Straws, used for drinking, 75
Strings, selection of to attain objective, 27 *seqq.*
Sultan, 4 ; with suspended fruit basket, 7, 19 ; with invisible objective, 20 ; uses right hand, 28 ; selects strings to

attain objective, *ibid.* ; uses sticks, 31 ; climbs on box to attain objective, 40 ; test, with two and more boxes, 45, 135, 138 ; uses ladders 47, 51 ; mounts keeper's shoulder, 49 ; climbs on door to reach objective, 54 ; swings on rope, 57 ; removes obstacle to attain objective, 59 ; jumps with pole, 69 ; feeds fowls, 85 *n.* ; miserable when isolated, 88 ; carries preserve-tin in mouth, 92 ; makes a noise with it, 94 ; makes implement from shoe-scraper, 101 ; from a branch, 103 ; unbends wire to make imple-ment, 111 ; uncoils rope, 113 ; fails to hang up rope, 117 ; empties obstacle-box, 119 ; fits bamboo rods together, 125 *seqq.* ; sharpens wooden splinters, 132 ; pushes keeper into erect posture, 142 ; invokes aid, 143 ; difficulties in building, 144 ; builds on stone heap, 155, and on cans, *ibid.* ; tries to use a ladder, 159 ; watches others build, 167, 170 ; makes double stick for Chica, 171 ; uses auxiliary stick, 174, 177, 183 ; pauses during experiments, 192 ; his stupidities, 196 ; moves string laterally to reach objective, 200 ; previous experience, 205 ; tested with wire-netting, 226 ; succeeds with "roundabout-way board," 232, 242 ; takes off nail, 245 ; pushes basket off rod, 251 ; pushes objective towards bars, 253 ; uses crooked stick, 258 ; confused by ladder, 260 ; indifferent to "form," 262 ; digs up objective, 279 *seqq.* ; behaviour when angry, 288 ; towards newcomers, 290 ; takes sides with human beings, 292 ; his obstinacy, 296 ; spirit of leaderhsip, 299 *n.* ; begs from Tschego, 301 ; copulates with her, 303 ; antics with his limbs, 312 ; afraid of a hen, 319 ; with a photograph, 325 *seqq.*

Table, used for reaching objective 47

Tank, opened by apes, 75

Tenerife, scientific station in, 4 ; ants in, 76

Tçrcera, 5, 50, 61, 70, 73, 139, 178, 200 ; with suspended fruit basket, 18 ; removes obstacle, 62 ; feeds fowls, 85 *n.* ; throws stones, 86 ; adorns herself, 92 ; her destructive-ness, 171 ; tested with "roundabout-way board," 239, 242 ; with ring on nail, 246 ; behaviour during Konsul's illness, 285 ; her jealous disposition, 296 ; friendship with Chica, 299

Teuber, E., 7, 101, 205, 299 *n.*, 306 *n.*, 313

Thorndike, American scientist, 3, 22 *seqq.*, 222, 278

Time, limitations of in chimpanzees' conceptions, 267 *seqq.*, 272 *seqq.*

Toy animals, chimpanzees' fear of, 320 ; dogs' and donkeys' attitude towards, 323

Tschego, 4, 26, 70, 73, 139, 213, 215 ; uses stick to reach objective, 31, 34 ; uses blanket for same purpose, 34, 38 *n.* ; climbs on door to reach objective, 56 ; removes obstacle, 62 ; method of digging, 78 ; dislike of lizards, 80 ; menstruation of, 81 *n.* ; uses stick as a weapon, 82 ; how punished, 86 ; her anger, 87 ; neglects Sultan, 88 ; makes a nest, 90, 93 ; adorns herself, 91, 92 ; her rhythmic marches, 92 ; treasures objects between abdomen and thigh, 95 ; in her lap, 96 ; makes a straw bundle for an implement, 196 ; breaks off branch for same purpose, 107 ; her perplexity when building, 147 ; fails with auxiliary stick, 176 ; tested with wire-netting, 226 ; with "round-about-way board," 238, 242 ; her need of peace, 292 ; mode of punishing her, 293 *n.* ; moral support of the others, 298.; friendship with Grande, 299 ; gives food to Sultan,

301; copulates with him, 303; dances, 314; admires herself in a pool, 318; sees her portrait, 325

Urgeschichte der Kultur, 323 *n.*
" Uswege," translation of, 11 *n.*

Volkelt, M., 164 *n.*
Vorstellungen der Tiere (Volkelt), 164 *n.*

Walking, methods of, 313
Wasmann, E., 190 *n.*
Wertheimer, M., 111, 116, 131, 163 *n.*, 225, 268
Wire-netting, tests with, 226

Yerkes, R. M., 268

Zeitschrift für angewandte Psychologie, 163 *n.*

International Library of Psychology, Philosophy & Scientific Method

Editor: C K Ogden

(Demy 8vo)

Philosophy

Anton, John Peter, **Aristotle's Theory of Contrariety** *276 pp. 1957.*
Black, Max, **The Nature of Mathematics** *242 pp. 1933.*
Bluck, R.S., **Plato's Phaedo** *226 pp. 1955.*
Broad, C. D., **Five Types of Ethical Theory** *322 pp. 1930.*
　The Mind and Its Place in Nature *694 pp. 1925.*
Burtt, E. A., **The Metaphysical Foundations of Modern Physical Science**
　A Historical and Critical Essay *364 pp. 2nd (revised) edition 1932.*
Carnap, Rudolf, **The Logical Syntax of Language** *376 pp. 1937.*
Cornford, F. M., **Plato's Theory of Knowledge** *358 pp. 1935.*
　Plato's Cosmology, The Timaeus of Plato *402 pp. Frontispiece. 1937.*
　Plato and Parmenides *280 pp. 1939.*
Crawshay-Williams, Rupert, **Methods and Criteria of Reasoning**
　312 pp. 1957.
Hulme, T. E., **Speculations** *296 pp. 2nd edition 1936.*
Lazerowitz, Morris, **The Structure of Metaphysics** *262 pp. 1955.*
Mannheim, Karl, **Ideology and Utopia** *360 pp. 1954.*
Moore, G. E., **Philosophical Studies** *360 pp. 1922. See also* Ramsey, F.P.
Ogden, C. K. and Richards, I. A., **The Meaning of Meaning**
　With supplementary essays by B. Malinowski and F. G. Crookshank
　　394 pp. 10th Edition 1949. (6th Impression 1967.)
Ramsey, Frank Plumpton, **The Foundations of Mathematics and other**
　Logical Essays *318 pp. 1931.*
Richards, I. A., **Principles of Literary Criticism** *312 pp. 2nd edition, 1926.*
　Mencius on the Mind. Experiments in Multiple Definition
　　190 pp. 1932.
Smart, Ninian, **Reasons and Faiths** *230 pp. 1958.*
Vaihinger, H., **The Philosophy of 'As If**
　428 pp. 2nd edition 1935.
Wittgenstein, Ludwig, **Tractatus Logico-Philosophicus** *216 pp. 1922.*
Wright, Georg Henrik von, **Logical Studies** *214 pp. 1957.*
Zeller, Eduard, **Outlines of the History of Greek Philosopohy**
　248 pp. 13th (revised) edition 1931

Psychology

Adler, Alfred, **The Practice and Theory of Individual Psychology**
　368 pp. 2nd (revised) edition 1929.
Eng, Helga, **The Psychology of Children's Drawings**
　240 pp. 8 colour plates. 139 figures. 2nd edition 1954.
Koffka, Kurt, **The Growth of the Mind** *456 pp. 16 figures. 2nd edition*
　(revised) 1928.